Phantoms of the Prairie

Phantoms
of the Prairie

The Return of Cougars to the Midwest

John W. Laundré

THE UNIVERSITY OF WISCONSIN PRESS

The University of Wisconsin Press
1930 Monroe Street, 3rd Floor
Madison, Wisconsin 53711-2059
uwpress.wisc.edu

3 Henrietta Street
London WC2E 8LU, England
eurospanbookstore.com

Printed in the United States of America

Library of Congress Cataloging-in-Publication Data
Laundré, John W.
Phantoms of the prairie : the return of cougars to the Midwest / John W. Laundré.
p.   cm.
Includes bibliographical references and index.
ISBN 978-0-299-28754-2 (pbk. : alk. paper) — ISBN 978-0-299-28753-5 (e-book)
1. Puma—Middle West. 2. Puma—Great Plains. I. Title.
QL737.C23L368   2012
599.75′24—dc23
2011043508

To Cecile and Lucina

Contents

Illustrations

Preface

The history of our treatment of predators is not a pretty one. In fiction and in real life we have treated them as vermin, as a scourge, as "weeds in the garden" to be pulled out and eradicated. Everywhere people of European decent have settled, they have entered what was an Eden of wildlife where prey and predator coexisted in an evolutionarily fine-tuned balance. This was the case in the "Serengeti" of North America, the Great Plains. In a historical blink of an eye, we reduced the vast herds of large prey and their predators, including cougars, to what is a landscape devoid of its animals and its wildness. For decades, the prairie landscape lay empty, haunted only by the memory, the phantoms, of past inhabitants. Now this is all beginning to change with one of those phantoms, the cougar, making a comeback. This is a book about what may be the promise, the hope that perhaps we still have a little bit of wildness in us, a bit of spirit to make room again for this phantom of the prairie to return to what remains of the great grasslands of our country. In analyzing whether we do or not, I have relied on my extensive knowledge of cougar ecology and drawn from the many excellent studies by other dedicated cougar biologists, to whom I give thanks. This book is not, however, meant to be a hard scientific work. It is one based on current scientific knowledge but also on my impressions of what is happening with cougars in the Midwest. I have written this book for everyone—scientist, citizen, politician—who is interested in what is happening with cougars in the Midwest. I hope the reader finds what I have to say interesting and informative but, above all, helpful in finding that little bit of wildness that is buried deep in our often-too-domesticated soul.

Phantoms of the Prairie

Introduction

WHY A BOOK ON THE COUGARS OF THE PRAIRIES? And why now? As little as ten years ago, such a book would be similar to one on the yeti or Nessie the Loch Ness monster. It would be a book about mythical creatures that spark the imagination but fall short in reality. It would be a book of maybes and what ifs, of raised hopes repeatedly dashed against the shores of objectivity and facts. Regardless of the many reported sightings of cougars, there was no solid evidence that they could be found in the prairie region. There have been large house cats, mistaken at a distance to be bigger than life. Large dog tracks in the mud or snow were too often mistaken by the inexperienced eye for the passage of a cougar. Most sightings have been reported by well-intentioned observers believing they have seen something out of the ordinary. However, some were downright hoaxes, pictures of cougars taken elsewhere but claimed to be from the region. All these, the honest mistakes and no-so-honest ones, diminished the creditability of people reporting them. Afraid of being labeled a crackpot, many hesitated, doubted what they were seeing. One might find more support in reporting a sighting of a UFO than a cougar. The official belief was that cougars no longer roamed the prairie region.

That cougars were there historically, there can be no doubt. As tame and "civilized" as the prairie region is today, before, it was a virtual wildlife paradise with millions of animals, "A Country So Full of Game" as James Dinsmore pointed out in his 1994 book of that title. Extending east to west from the western edges of more forested states such as Wisconsin, Illinois, Missouri, and Arkansas to the foothills of the Rocky Mountains and north to south from Canada to Mexico (fig. 1), this vast region was a sea of grass. Roaming this sea

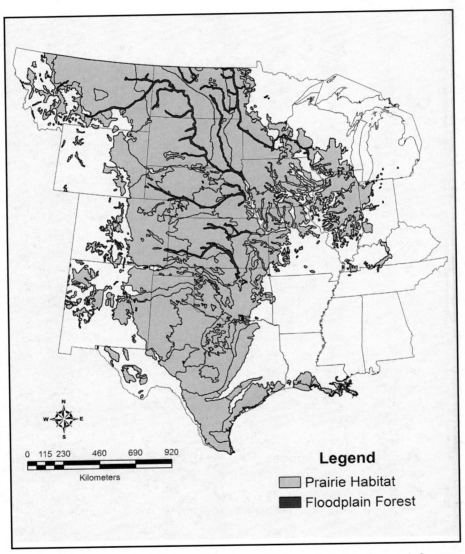

FIGURE 1. Extent of prairie ecosystem in the United States. (Map based on work from Kuchler 1964 and digitized by the U.S. EPA.)

were herds of ungulates—bison, elk, deer, and pronghorn—along with their predators: wolves, coyotes, bears, and . . . yes, cougars. The numbers and diversity of wildlife in the region rivaled the plains of Africa. What happened within a relatively brief period of time, less than ninety years, is one of the wildlife tragedies of the world. From the time the epic expedition of Lewis and Clark in 1804–6 opened up the Western frontier to 1890, when that frontier could no longer be recognized, an unfathomable amount of wildlife was killed, often in a systematic, government-sponsored manner (fig. 2).

Unfortunately, the prairie was ideal for supporting not only a sea of grass but also amber waves of grain. Population and political pressures and a sense of manifest destiny encouraged, indeed mandated, the destruction of native American cultures and the seemingly endless herds of bison, elk, deer, and pronghorn along with their predators—wolves, bears, and . . . yes, cougars. Only the wily

FIGURE 2. Example of massive destruction of wildlife on the prairies where the killing of millions of wildlife, especially bison, pictured here, was sanctioned by U.S. government policy to make room for pioneers settling the region. (Photo courtesy of the Burton Historical Collection, Detroit Public Library.)

coyote persisted. Within an ecological blink of an eye, the prairie region was deprived of its wildlife heritage and its identity, becoming referred to as the "corn belt" or "wheat country." With this loss, modern-day inhabitants of this region have grown up without knowing or understanding what was there before. Few midwestern farmers appreciate the significance of the fact that millions of bison roamed over their now plowed and tamed cropland or that wolves once howled where now only can be heard the neighbor's dog. For tens of thousands of years, a whole living, functioning ecosystem, including humans, lived on a thin mantel of prairie grass supported by some of the richest soils of the world. Today, it would be a misnomer to call it an ecosystem, especially a self-sustaining one. By all indications, it is a system in decline, where row crops, too intensively cultivated and fertilized, slowly drain the life from the soil. Only if the modern inhabitants of this region take the time to visit the museums or read the history books would they know what was lost. However, they can never truly appreciate the extent or the magnitude of what the prairie ecosystem was—a truly wild and free land. That wildness and freedom, unfortunately, is perhaps lost forever.

With the destruction of the prairie wilderness, pioneering ancestors may have ensured safety and prosperity for their descendants but deprived them of the heart and soul of the land that drew them initially. Where their pioneering ancestors woke to the howling of distant wolves, residents now awake to the sound of diesel engines, on the farms or on the streets and highways. Where early residents could ride freely across the prairie, thriving in the liberty of these open spaces, residents of today are confined by fences and roads that divide the landscape into a checkerboard of square-mile sections. An artificial geometry was forced on the landscape, depriving the region of its original unbroken personality and its open spirit. From the writings of John Muir (1912) we know that wilderness nourishes the soul. The nourishment that fed the souls of the pioneers on the plains died with the last free roaming bison, the last wolf, the last cougar. From that time on, the region became safe but predictable, no surprises, no excitement, no "wildness."

But today, things are different. There is an excitement in the air. Incredible things are happening in this area, in what many considered to be the most settled region of the United States. People are being shaken from their post-pioneer slumber and complacency from living in a land long devoid of its wildness. Living in a land where the most exciting things to happen are state fairs and high school championships. Living in a land where today, however, one of the greatest ecological events in a lifetime is unfolding before our eyes. Cougars, pumas, mountain lions, panthers—these are no longer figments of peoples' imaginations, fleeting sightings of something that possibly, maybe, could be, but probably wasn't something real. The cold, hard facts, usually the ally of the doubters, are now on the side of the crackpots, the often belittled believers in what they saw. There is no doubt that cougars have been sighted in the prairie regions of Middle America. There are not just the occasional often-suspect sightings or blurred tracks; there are pictures, and, unfortunately, there are dead animals in hand. After a hiatus of more than a hundred years, there is undeniable proof that cougars have been sighted in every midwestern state of the prairie region.

From a wildlife ecologist point of view, these are indeed exciting times. In a world where large predators are declining and disappearing from their current range, to have one actually expand into its former haunts engenders admiration for its tenacity. It brings hope for the return of a little wildness in a tamed landscape. This is especially the case for a wildlife ecologist like me, born and raised in these tamed lands. During my childhood and young adulthood, large predators were animals only to read about, to dream about, to see on *Wild Kingdom* in some faraway land. To see or work with them meant pulling up stakes and moving to these wilder places where they still roamed. And that is what I did; like so many young students interested in large predators, I moved West where cougars and grizzly bears still roamed, in the hopes of studying them, in the hopes of seeing them.

My hopes and dreams were fulfilled in the mountains of southern Idaho and northwestern Utah where I got the once-in-a-lifetime opportunity to study cougars. However, this area was not the pristine wilderness where earlier cougar biologists had worked. It was an area of small mountains surrounded by range and farmlands. It was right in the middle of the home of those famous Idaho potatoes. From the tops of my mountains, I could see fields of potatoes as well as the Great Salt Lake Valley with its increasing suburban sprawl. People lived and worked right next to and within areas that cougars roamed. They went for picnics and hikes up into the mountains right in the midst of cougar habitat.

It is in this seemingly tame landscape that I spent sixteen years studying the ecology and behavior of America's big cat. Besides learning a lot about cougars, I learned about people living with cougars in their back yards. And what I found reshaped my views, my fears, and my hopes regarding cougars. Unlike the fear and apprehension held by people who for generations never saw cougars, western folks were accustomed to them. They did not fear them; in fact most times they hardly knew they were around. Cougars rarely bothered the cattle in this region, and without concern ranchers and hikers alike roamed the mountains where cougars lived. Attitude surveys supported this lack of concern about personal safety; people of all walks of life liked the idea that cougars still roamed the woods. Even ranchers accepted their presence, expressing only their concern to be able to respond to "problem" individuals, which were few and far between. There was no fear, no hysteria, as we too often see elsewhere. There was a peaceful coexistence maintained mainly because of the exemplary behavior of the cougars.

Having brought my naive impressions, instilled from growing up without large predators, I was cautious at first but quickly converted to the western way. Although I continued to respect this powerful sleek predator, I no longer feared it. In my work, I gained close-up experience with cougars in a multitude of situations. We treed them with dogs, we climbed up the trees to take pictures; for one of the houndsmen who helped me, his idea of a telephoto camera lens was to climb as close to the treed cat as possible and then extend his arms even closer to snap the picture with his simple point-and-shoot camera. We walked in on cougars and chased them away from deer they had killed. We walked in on females with small kittens, grabbed the kittens, and weighed and measured them as the females stayed nearby out of sight and watched. We did all those things that, as a young boy, I had heard would incite the wildest and most ferocious reactions in these killer beasts. What was more dangerous than to get between a mother and her offspring? But yet, I survived to write about it. I still have both my hands and all my fingers. I remain scar-less, never once being threatened, let alone attacked, by those cornered cats or those distraught mothers. In fact, the more I worked with them, learned about them, the more I came to see that cougars were probably one of the most timid of large wildcats out there. Though they kill for a living, as we do directly or indirectly, they are no more vicious or ruthless than we are, at times probably less so. Just as we who hunt do not kill a deer out of anger and hatred, neither does the cougar; it is just doing what it is supposed to do. When not acquiring its food, a cougar is shy, bordering on a recluse. It is rarely seen and does not seek out aggressive encounters with other predatory species, including humans. Even when I knew where a cougar was because of its radio collar, I rarely had the opportunity to actually see it. Living up to their name as the "ghost of the Rockies," they silently slipped away rather than confront. Because of this timorous behavior, especially toward humans, it amazes me that cougars do occasionally attack and kill people. I am convinced, however, that those attacks are never out of anger or viciousness but more from a misdirected hunting instinct, often the result of a case of mistaken identity.

My work with cougars over those and subsequent years reshaped my views. Facts replaced beliefs born from unfounded stories and fears. One of the greatest transformations was that I came to learn and appreciate the role of predators, especially large ones, in nature. As many still do, I originally viewed predators as "weeds in the garden," destructive animals that harmed the ecosystem, to be removed at all costs.[1] I, too, killed my share of them because it was the right thing to do. However, working with cougars, I came to see them not as weeds but as the tenders of the garden, the shepherds of the wild flocks, and thus the guardians of ecosystems. Without these vigilant custodians working 24/7/365, wild herbivores would, and have, become the true weeds in the garden. Moving unrestricted and unafraid, deer and elk eat where they want and what they want. They strip the environment of their preferred foods, leaving less palatable and often exotic species, weeds, to dominate over a simpler and less productive landscape. Predators maintain the order in the garden; they keep herbivores in their ecological rows. They do this not so much by killing them but by scaring them.

The landscape-of-fear model my colleagues and I proposed provides us insights on how predators tend the garden.[2] Any predator, such as cougars, can't be equally efficient or lethal in all habitats. They have their specialties and their weaknesses. For example, cougars are best at catching deer along forest edges, leaving deer relatively safe in open areas.[3] This spatial variation in a predator's lethality creates a landscape where prey have to balance the desire to eat with the fear of being killed: thus, the landscape of fear. In the areas where predators have the advantage, are more lethal, prey fear to go. This creates refuges for those plant species preferred by herbivores. There, they can survive and grow, preserving the plant diversity in an area. Remove the predators, you remove the fear, and you remove the refuges.

These things and more I learned as I followed cougars around the mountains of Idaho and Utah. Most of what I write later in this book comes from those experiences and research. I had to travel far to the west to have those experiences, but now, right in my former backyard, these magnificent cats are returning!

To see and participate in this ecological phenomenon is truly a second once-in-a-lifetime opportunity for me. Knowing what I do about cougars, I, like so many others, watch with wonder and anticipation as this ecological drama unfolds. We embrace the possibility that a little wildness will return to these

tamed lands. We too can sense the excitement others farther west have, knowing that cougars are out there. Additionally, those of us trained in ecological
theory know the benefits of having top predators in an ecosystem. We appreciate their role in controlling the unabated foraging of large ungulates. Such foraging, akin to domestic cattle, endangers rare plants and encourages invasions
of exotic species. We have seen and studied the cascading beneficial effects
of the return of wolves to Yellowstone National Park. For us, the return of a
top predator to the prairie ecosystem is viewed as the start of an ecological
healing process.

For others, the thrill and joy of seeing cougars return to the heartland of the
country is replaced with fear and apprehension. Long having lived without
cougars, or other large predators, many people don't know what to expect. Their
fear of the unknown is fanned by our cultural views of predators in general,
and large ones in particular, as menaces to humans and their endeavors. Our
books and tales reflect and reinforce this view. They grew up with Little Red
Riding Hood and the three little pigs; predators are cunning, they are bloodthirsty, they are bad. Because of this, people see the cougars' return as a threat,
to themselves, to their livestock, to their game species. They wonder, will we
have to live in constant fear that these large cats not only will prey on deer but
on us as well? They argue, our ancestors waged a national campaign to remove
these threats for the good of society and future generations, so that we would
not have to deal with them. Why would we want to see their return?

Two conflicting views for sure. Where does the truth lie? Is the possible
return of cougars to the plains region a blessing or a curse? Will we be able
to coexist peacefully with cougars or will we live in constant fear? Is the risk
just too great to even entertain the idea of their return to this region? Much of
all this uncertainty is because we lack information on cougars in the plains
regions. True, there have been many books written on cougars, which cover
various aspects of cougars in general,[4] cougars in specific regions,[5] cougar-
human interactions,[6] and many more, including a recent, all-inclusive review by
renowned cougar biologists.[7] Added to this is the growing body of scientific literature on research done on cougars in the Western United States and Canada,
Florida, Mexico, and Central and South America. There have even been publications on cougars in the Eastern United States, primarily on the "potential"
of their return.[8] A lot of information exists, but the two aspects in common in
all the U.S. books and studies is that they are from areas of forested habitat or
mountainous terrain, much of which is forested. Conspicuous (because of its
absence) is information on cougars in the midsection of the United States, the

"Heartland," the Great Plains, the prairie regions. The simple reason for this lack of regional information is because cougars were removed from the area before the science of studying wildlife began. All that exists are records, mostly records of the removal of cougars from the plains region along with pictures of proud hunters killing the last of these vicious beasts (fig. 3).

Although more than 90 percent of this region today is devoted to crop production or intensive cattle grazing, at one time it comprised one of the largest single native habitat types in North America, the grasslands or prairies. Officially this region is referred to as the grassland biome and is a product of unique temperature and moisture conditions that favor grasses over more woody vegetation. Because of its extensive area, the original grassland vegetation was not all the same. We commonly recognize three major East-West divisions: tallgrass, midgrass (or mixed-grass), and shortgrass prairies. Before conversion to the American breadbasket, this region was the domain of the vast bison herds as well as elk, deer, pronghorn, wolves, and cougars. Today, deer can still be found in most of the region, and pronghorn still occur in isolated areas. The bison

FIGURE 3. Cougars were commonly hunted and killed without limit or control in the early days of settlement of the prairie region. (Photographer unknown, photo courtesy of U.S. Fish and Wildlife Service.)

herds are long gone, along with the wolves and elk, but the cougars seem to be returning. In recent years cougars have crossed grassland habitat to re-colonize the Black Hills of South Dakota and the Badlands of North Dakota. Verified sightings of cougars are increasing yearly in states such as Missouri, Nebraska, and as far east as Wisconsin and Illinois.

So after an absence of more than a century, cougars seem to be on the move back into this vast region, where we know the least about them. There are just scattered accounts of their pre-settlement distribution, and we know next to nothing about their habitat preferences and ecological role in those times. This makes it difficult to evaluate their prospective for survival as viable populations in this region. Maybe the prairie region has become too settled to support cougars, and the discussion of their returning is moot. If it is possible for cougars to survive in the Midwest, where is this most likely to occur? Are there enough habitat and prey available for them to make a living? What ecological role would they play in this altered ecosystem? Would their presence, anywhere, necessarily lead to conflicts with humans? What is the actual risk we would face with the return of cougars? How do we deal with "problem" animals that show up in our suburbs? Even if cougars can't return to stay in the plains, maybe they can pass through to points farther east.

The plains region is the largest area of potential eastern movement of individuals to possibly re-colonize those forested and mountainous areas. Can the cougars do it or are the extensive areas of cropland too impossible of a barrier to cross? If they can, what are the most likely travel routes they would take? These and many more questions were only hypothetical a short time ago, questions for a mental game of "what if" that animal ecologists and naturalists played. Now with cougars moving out into the prairie region, they become questions that urgently need to be answered. Answers that hopefully will help the scientists, the public, and elected officials make rational decisions and policies regarding this ecological phenomenon. However, without basic information on cougars in the prairies, these answers will be hard to come by. What will take their place are conjecture and supposition based on our cultural biases and fears regarding predators in general and cougars in particular. Unscrupulous politicians already are using these biases and fears to their political advantage, whereby creating a polarized, confrontational mood all too familiar in modern-day politics. Unfortunately, such polarization will lead to the social atmosphere that initially caused the demise of the cougars and the prairie ecosystem and surely will not resolve the modern-day situation in a reasonable and rational manner. To prevent this it is imperative that we have some scientific basis for

evaluating the role of cougars in the original prairie ecosystem and what it may or may not be in the current landscape.

This book is an attempt to fill the information gap we have on midwestern cougars. I look at the history of cougars in the prairie regions and attempt to reconstruct what cougar populations were like before settlement. I present when cougars were extirpated from the various states and where. I place special emphasis on where the last animals were killed, as these likely represented last refuge areas that might still provide adequate habitat for cougars. Based on what is known about cougar ecology and behavior, I try to surmise what their potential ecological role might have been in the original prairie biome. This supposition is based on not only my research efforts but also on the work of many excellent cougar biologists who came before me and those who are continuing the work. I acknowledge all their fine contributions and cite much of their work in the reference list for the reader to learn and enjoy. Once I have reconstructed the cougar's past role, I then examine the evidence of re-colonization of this region by today's cougars. From this examination I assess the potential of each state in this region to maintain a viable population by identifying where those areas are most likely to be and how many cougars we might expect. I summarize the potential for cougars to move across this vast region to more eastern points. Even if there is limited room in the former prairie environment, if cougars could safely travel through these lands, they could repopulate the vast forested regions east of the Mississippi. I try to view the landscape through the eyes of the cougar by taking a virtual journey across the modern landscape to identify most likely corridors and most probable barriers to cougar movements.

Last, I evaluate if there is room sociologically and politically for cougars in the Midwest. Even if the biology says the prairie region can support cougars, do people of the Midwest have the social and political will to accept them? Can people of the Midwest, like their western counterparts, tolerate and co-exist with cougars? Are they willing to accept a little risk in their lives? Or have they lost the wildness of their ancestors, the pioneer spirit to live with nature on its terms, just a little? If they are willing to accept cougars back, just how much risk are we talking about? What impacts will cougars have on our game populations and livestock? What are the chances of humans being attacked by cougars? How tolerant should we be? How should we deal with cougars that show up in suburban/urban areas? The answers to these questions are important to help determine if viable cougar populations can exist together with current and future human populations. If the citizens of this vast region are willing, then there might be room for the cougar's return. If they are not, then the Midwest will play the role only as a population sink where western cougars go but are never heard from again, except maybe dead on the nightly news.

This is an ambitious plan, but because of my years of experience with cougars and with the published efforts of so many other cougar researchers, it is not an impossible goal. I have the added advantage of having lived in the Midwest; I am familiar with the types of modern-day landscapes that face these pioneering cougars and what the historic landscapes were like where their ancestors lived. So come along with me and the cougars on this adventure. Ecologically it will be an interesting and enlightening journey, and hopefully along the way you will learn more about America's lion and its role in the ecosystem.

Before I begin, I need to clarify the exact area I am talking about. If I talk about the Midwest or the prairie ecosystem or the Great Plains, it can be confusing because each term covers different areas. The term "prairie ecosystem" is the broadest definition, and it covers a wide swath of land from Mexico well into Canada. My intention is not to cover all of this area but to limit my discussion primarily to much of the prairie habitat in the United States. What areas and which term, then, should I use? The term "Great Plains" covers ten states from Montana and North Dakota to the north to New Mexico and Texas to the south (fig. 4). All these states have prairie habitat but officially the Great Plains does not cover *all* the prairie ecosystem in the United States (see fig. 1). So in itself, the term "Great Plains" is not adequate. If I use the term "Midwest," it includes twelve states from North Dakota to Kansas in the west to Ohio and Michigan in the east. This term does include many states with prairie habitat that are not listed as the Great Plains, but it also includes states such as Michigan, Indiana, and Ohio that have little if any prairie habitat. So what term do I use when defining and discussing the area of interest? Probably "Midwest" comes closest to my area of concern, if I stipulate that this refers to states that historically had significant prairie vegetation and also cougars. This then, would exclude Indiana, Ohio, and Michigan. However, the term "Midwest" excludes Texas, New Mexico, Colorado, Wyoming, and Montana. Though these states do have grassland habitats, they still have resident populations of cougars within their borders, and I do not discuss these areas directly. If I add the requirement that the cougars had been extirpated from the states after settlement by eastern immigrants, this then would exclude these states. What remains are the ten prairie states that I do want to consider directly: North Dakota, South Dakota, Minnesota, Wisconsin, Iowa, Illinois, Nebraska, Kansas, Missouri, and Oklahoma. To this list I would add one additional state, Arkansas. Although not considered either a prairie or Midwest state, as explained later, it does play an

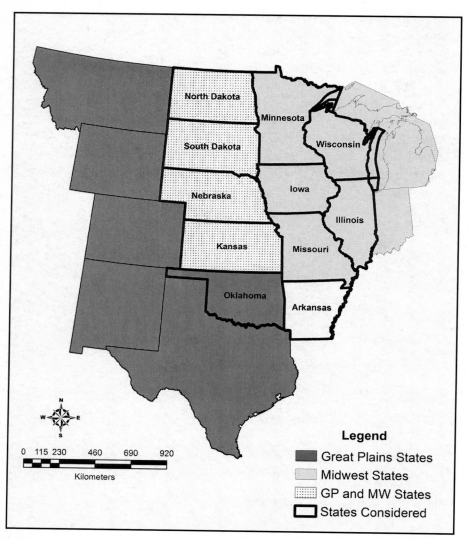

FIGURE 4. Because different areas are covered by the terms "Great Plains" and "Midwest," this map shows those states referred to as "Midwest" in this book (outlined in black). (Map by author.)

important role as a "border" state relative to the potential re-colonization and survival of cougars in the ten prairie states I will consider. Although now a mix of terms and states, to keep things simple, I will refer to these eleven states as the Midwest states. I will occasionally still refer to the prairie ecosystem or the plains region, which should be understood to be synonymous to the Midwest as I have defined it here.

Having defined my area of interest, we are ready to start our journey regarding the past, the present, and, hopefully, the future of cougars in the midwestern prairie ecosystem. The first place to start is in the past. Before there were houses, before there were fields, before the Midwest was "civilized," where were the cougars on the prairies? By looking at the pre-settlement records and accounts we can begin to develop the picture of where and how cougars lived in this vast sea of grass.

I Pre-Settlement Records and the Demise of Cougars on the Great Plains

BECAUSE THE INDIGENOUS PLAINS PEOPLE did not leave much of a recorded history of their lives in the region, very little can be surmised about cougars from them. What is known is that cougars were present on the plains because they do play roles in some of the oral history handed down through stories and rituals. But even though cougars did figure in indigenous peoples' lives, they did not appear to take on as big a role as other species such as wolves and bison. On the eastern edge of the prairies, early cultures included cougars or panthers in their effigy mounds, but little is known of their significance. In the heart of the prairie lands, the bison was the center of Plains cultures because it provided the food and many products they needed. Wolves, because of their obvious presence, also dominated the indigenous society. The scarcity of cougars woven in native cultures could have been because of the secretive nature of cougars or because they were not very commonplace in the everyday lives of early plains people. It is known that Plains people hunted cougars and used their skins as clothing and ornaments. R. G. Thwaites's extensive collection of early accounts mentions cougar skin as a saddle blanket for horses.[1]

In the Plains people's symbolism, cougar fetishes and totems were symbols of leadership, power, courage, and decisiveness.[2] However, there are few stories and legends of cougars, especially in comparison with other species such as wolves and coyotes.[3] No particular threat or danger seems to have been attached to cougars, and there appears to have been a peaceful coexistence between humans and cougars over the millennia before the arrival of European explorers.

The first Europeans to enter the vast grasslands of North America actually came from the south out of Mexico. The Spanish explorer Francisco Vásquez de Coronado led an expedition in 1540–42 into the prairie region to what is now Kansas. A fairly detailed account of his travels by Pedro de Castañeda de Nájera has survived.[4] This account primarily talks about the various indigenous peoples they met, their interaction with them, intrigues among the various personalities involved in the expedition, and the expedition's obsession with finding gold and silver. Because the Spaniards were more interested in finding riches, their accounts rarely include detailed descriptions of the native fauna and flora. This account is no exception. Apart from a brief mention of the "cows" (bison) and wolves, few other accounts of wildlife on the plains were listed, but they do mention that the wolves were "numerous" (*abundante*). This is noteworthy not because of their crude abundance estimate but because they took note of a prominent predator. Cougars were known by the Spanish in Nueva España, now Mexico, and we could assume that if they had seen cougars on their expedition, it would have also been worth noting. The lack of reference to cougars might be taken as an indication of the rarity (or secrecy) of this animal in the plains.

Following Coronado, there were various minor excursions into the plains region, primarily the French who came from the east in the mid- to late seventeenth century. These journeys by people like Marquette, Joliet, and others did not extend very far into the plains area and rarely mention any wildlife they might have seen. From 1766 to 1768, the Englishman Jonathan Carver ventured into the beginnings of the prairie areas of the upper Mississippi River. He traveled two hundred miles up the St. Pierre River (modern-day Minnesota River) where he spent five months with the Naudowessie of the Plains tribe.[5] Like many others, Carver devoted most of his narrative to the customs of the native people, but he did devote one chapter to the "Beasts," which are found in the interior parts of North America. Here, he talks about the "tiger," which presumably is the cougar. He describes them as smaller than those of Africa and Asia, free of spots, and not "so fierce and ravenous" as the others. He reports seeing one on an island in the Chipeway (Chippewa) River that he was able to view at close range. He observed that it did not seem apprehensive of his approach nor did it have any "ravenous inclinations." His final comment was that they were seldom seen in "this part of the world."

A. P. Nasatir in 1990 compiled and published an impressive number of documents from 1785 to 1804, written by a wide range of persons who had traveled various areas along the Missouri River. He consolidated them into two volumes to document the history of Europeans in this region before Lewis and Clark made their epic journey. Most of these documents consisted of letters written back and forth from people living and working in the "frontier" to either family members or business associates living back East. Although these volumes provide a rich source for learning about human life on the plains, few of these documents dealt with wildlife, and again there was no mention of cougars.

The Lewis and Clark expedition (1804–6) was unique in that its intended goal was to document the various natural resources that America now had in possession from the Mississippi River to the Pacific. They kept detailed accounts of their journey, which included many references to the wildlife they observed. They spent a considerable amount of time in the plains region during their trip out and on their return. This included spending the winter of 1804 in the middle of North Dakota at Fort Mandan. Although they encountered a great variety and abundance of wildlife species, there was little mention of cougars. Paul Johnsgard, in 2003, published a book in which he lists all the prominent animals and animal species he found mentioned in Lewis and Clark's writings. The cougar was mentioned only two times in the extreme western edge of the Great Plains in Montana. Again, this attests to the rarity and secretiveness of cougars.

After Lewis and Clark, the West began to be crossed by adventurers, hunters, and trappers. Many passed quickly over the plains region for the diverse trapping opportunities in the mountainous regions. At that time, few spent time in the plains and had little interest in the vast herds of bison. In most cases, few wrote detailed accounts of what they found. Between 1833 and 1843, though, there was one noted exception that provides us with some information on the wildlife encountered in the plains areas. Francis Chardon was the "Factor" or overseer from 1834 to 1843 at Fort Clark along the Missouri River north of modern-day Bismarck, North Dakota. In his diary (1834–39), Chardon left us a detailed account of daily life at the fort, which included estimates of various wildlife species killed by the fort's inhabitants and the local natives.[6] Obviously, bison and beaver were the most noteworthy; but he did also detail the number of "wolves" (likely a mixture of wolves and coyotes) and foxes he trapped. During that time, there was no mention of cougars being killed or sighted. Because cougar fur is not very desirable, early trappers and hunters would not have made a special effort to kill them. However, it would seem that cougars are rare enough that if they had encountered any, some note of it would have been made. Again, more evidence that cougars were either rare or secretive, or both.

After Lewis and Clark, most trappers and settlers that passed through the plains areas did so quickly, heading toward rich hunting grounds and farming areas to the west. Few accounts have survived, and there is no mention of cougars. R. G. Thwaites, in his impressive thirty-two-volume set of accounts of early (1748–1846) travelers in the West, included the account of the naturalist Thomas Nuttall and the translation of the accounts of Prince Maximilian of Wied, Germany.[7] Most of these rare accounts of early travel into the plains region dealt primarily with the native people encountered, but some, such as those of Nuttall and Wied, provided more descriptions of the flora and fauna. Of the animals mentioned, again, wolves (coyotes and wolves) and bison dominated. When mentioned, the reference to cougars (panthers) was often brief; for example, in the De Smit letters, cougars were mentioned as common along the mouth of the "Yellow Stone" river.[8] In the notes of the S. H. Long expedition, panthers were listed with various other species known to be in the region.[9] Nuttall remarked that panthers were common along the forested banks of the Red and Arkansas Rivers.[10] Prince Maximilian presented the most extensive listing of cougars, primarily referring to seeing their skins being used as saddle blankets or arrow quivers.[11] Even so, he speculates that the plains natives probably obtained the skins through trading with more mountainous tribes. The picture of cougars on the plains areas in pre-Euro-American settlement is one of scarcity, amplified by the secretive nature of the species. We can assume that cougars were there but not much more regarding habitat use or abundance can be determined.

To get some idea of the occurrence and habitat use of cougars in the plains region, we turn our attention to later times when finding and killing a cougar made big news. What follows are the state-by-state accounts of cougars being killed and eventually eliminated from the plains region.

As with so many wildlife species in America, the killing of predators was often big news. This is especially true for the cougar, which was considered to be a bloodthirsty killer of innocent wildlife and domestic stock. To kill a cougar was to do a social service akin to getting rid of the town bully or killing the local desperado. A person was often viewed as a local hero for slaying such a terrible and destructive beast. Between Lewis and Clark and the elimination of the last vestiges of wilderness on the Great Plains, the eradication of predators was a top priority, and men did it with a fervor bordering on the religious. One can see on the faces of the men posing with the dead corpses (see fig. 3), they truly believed they did a good and sacred deed. The taming, and thus the destruction of the prairie ecosystem, was taken seriously and was accomplished in an astonishing rate. By 1890, only eighty-six years after Lewis and Clark, the U.S. Geological Service declared it could no longer distinguish where the frontier began; which meant that the "frontier" no longer existed. In the eighty years after Lewis and Clark, the great herds of bison disappeared. So too did the abundant herds of elk and pronghorn, killed by the tens of thousands of pioneers crossing to the West and by the settling and plowing of the seemingly endless grasslands. Some species fared better; they persisted in the wild a little longer, and in some areas, the cougar was one of them. Maybe because of its rarity or secretiveness, cougars in many current prairie states persisted to the late 1800s and even into the early 1900s. Eventually they were all hunted down and killed. What follows is a record of the demise of the prairie cougar, state by state. Much of this information comes from records compiled by Young and Goldman in their book *The Puma*.

Recorded Occurrences of Cougars in Plains States

Arkansas

Although Arkansas is not a prairie state, as I explain earlier, as a border state it does figure importantly in cougar–prairie dynamics historically and, hopefully, in the near future. As background for future discussions, I present the information on when and where the last cougar was killed in this border state. Originally cougars were well distributed in the state, particularly along lower breaks of the Mississippi and Arkansas Rivers. They were also reported along the Louisiana border in Chicot County. They were still in the state in 1880 with the report of three cougars being killed in one day (possibly a family group) in Crooked Bayou, Chicot County, in spring 1887. No report of killing the "last" cougar was recorded, but in 1920 the report of one in Franklin County was sufficient to initiate an investigation. From that time to the present, occasional sightings of cougars have been reported, indicating the persistence of the species in the state but so far not of a sustainable population.[12]

Illinois

Cougars were once considered common in forested areas of Illinois, being reported from various counties in the early 1800s. Hoffmeister lists several reports of cougars being killed in the early 1800s. One of the earlier reports is of "panther" being listed as a species that could be found along the Illinois River in 1818–19. At least two creeks were named "panther creek" to commemorate the killing of cougars, one in Hancock County in an unknown year and another in Christian County in 1825. In 1823 people reported seeing a cougar in a forested area in Vermillion County and Madison County, and cougars were reported present as late as 1850. Later records of cougars include one being sighted about

1858 near Brownsville and one being killed in Macoupin County in 1840. The last one officially reported killed in the state was east of Thebes in Alexander County around 1862.[13] From that time on, only unsubstantiated reports of cougar sightings exist, one in Pope County in 1905,[14] and most assumed cougars had been exterminated from the state by 1870. Based on the records that exist, cougars in Illinois were found mainly in forested areas and along the rivers.

Iowa

The cougar was never considered common in Iowa, but there is one reference stating that it was formally common.[15] Young and Goldman, however, did not put much value in the accuracy of this source.[16] Dinsmore reports various cougars being killed in the mid-1800s.[17] According to definite records, there were six killed in Clayton County in 1838, one killed in Delaware County in 1845, one in Montgomery County in 1851, one or two seen in Madison County in 1861, one seen in Adair County in 1864, and one killed near Cincinnati in Appanoose County in 1867. None of these records had information as to what type of habitat the cougars were found in. However, other reports presented by Dinsmore do describe cougars being encountered in forested areas.[18] One of these was a cougar treed by a family dog in the woods of Winneshiek County in 1843. Another one was in 1855 where a boy was reported attacked by a cougar along the Boyer River in Crawford County. A third, undated reference is of a small cougar being killed along the Rock River in Sioux County. Dinsmore provides a map indicating the counties where records of cougars existed and for some, the years of the last known animal.[19] Based on what he presents, as late as the 1800s cougars were widely distributed across the state. It is unlikely, though, that they ever occurred in high numbers, except in the eastern part where there was extensive forested areas.[20] The last one killed in the state was the one in 1867 from Appanoose County in south central Iowa.

Kansas

In the late 1800s, seven specimens were reported from Harper, Barbour, and Comanche Counties in Kansas.[21] Three of these animals were killed, and one kitten captured and kept as a pet. J. R. Mead (1899), cited in Young and Goldman, recorded that they were rarely seen in the state but occasionally found in central Kansas and common along the southern border of the state. Most of these sightings were in the forested habitat of the streams and rivers that passed through the upland prairie grasslands. In 1859 one large cougar was killed in a heavily timbered bend of the Solomon River, a few miles from its mouth in

Saline County. Mead also reported seeing one in 1865 on White River in Butler County near the modern town of Towanda. In 1864 he saw a large male in the prairie about three miles south of the junction of the Medicine Lodge and Salt Fork Rivers, near the Great Salt Plain close to a ravine. J. A. Allen (1871), cited in Young and Goldman, reported the cougar rare on the prairies because of the lack of forest cover. He stated it was principally found along timbered waterways and breaks. C. W. Hibbard in 1943 listed the eastern cougar (*Felis concolor couguar*) as common in eastern Kansas at the time of early settlement but by then extinct. In western Kansas the Rocky Mountain cougar (*F.c. hippolestes*) occurred in the western part of the state. William Applebaugh and J. H. Spratt killed the last cougar in the state on August 15, 1904, in Ellis County, in the center of the state. One taken in late 1930s at Valley Falls was considered to have been a transient.

Minnesota

The first European, Etienne Brulé, entered what is now Minnesota in 1622 and was followed by various others including Josiah Snelling, who built the first fort in the region in the early 1800s. Over that time, increasing numbers of Americans settled in the region, but little was recorded regarding cougars. Carver traveled through the region in 1766–68 and reported seeing a cougar on an island in the Chipeway (Chippewa) River.[22] The Chippewa flows through west-central Minnesota, which was originally a mixture of prairie and open oakland vegetation. Roberts, as cited in Young and Goldman, reported them as never being common, and Suber in 1932 observed they were only rarely reported by early explores. There were few reports of cougars being killed, and the last one was in 1875 in Sunrise, Chisago County.[23] This was in forested habitat in the central eastern part of the state near the state line with Wisconsin. No references were found of cougars in the southwestern prairie area of the state.

Missouri

Cougars were reported as being part of the fauna around St. Louis around the 1800s.[24] In 1934, Prince Maximilian noted it as being occasionally seen along the Missouri River below Liberty in extreme western Missouri.[25] Cougars continued to persist in the state with the last recorded one being killed in the fall of 1888 along the Current River in south Central Missouri.[26]

Nebraska

No earlier reports exist than that of Aughey in 1880, where cougars were reported a few times on the Niobrara and Loup Rivers near the sand hills region of Nebraska.[27] J. K Jones reviewed records of cougars in the state and reported that originally cougars "occurred throughout the pine ridge area in the northern part of the state and occasionally ranged south and eastwardly along many of the larger streams." He reports two "definite" records; the first was a skull of an adult female with kittens from Hay Springs, Sheridan County, that were killed in 1884. The second was a mounted specimen that was killed in 1880 at the mouth of a canyon near Valentine, Cherry County. Jones also reported that based on observations by others, cougars were reported from southern Holt County as late as 1899 and that one was seen along the Republican River, south of Franklin, around 1888.[28] Consequently, it appears that cougars were absent from the state from around the turn of the century to the first confirmed report of one in the state in 1991 (see chap. 5).

North Dakota

Cougars were considered to have been seen throughout North Dakota but not overly abundant.[29] Historic records compiled by the North Dakota Game and Fish Department report that cougars in the 1800s could be found along the Little Missouri River, North Dakota Badlands, Killdeer Mountains, and the Missouri River Breaklands.[30] Specifically, they were rare east of the Missouri River valley but more common in the western, rougher areas. Cougars were reported as being extinct in the state by 1914 with the last record of one being killed in 1902 along the Missouri River near Williston in western North Dakota.[31] However, a sighting of a cougar was reported on June 16, 1943, eight miles south of Tagus, Mountrail County, west of Carpenter Lake. Because of its proximity to Montana with existing populations farther to the west, it is likely that cougars occasionally dispersed to North Dakota through much of the twentieth century. Thus from 1902 to the current reestablishment of cougars in the state (see chap. 3), there is some doubt if cougars ever were totally eliminated from the state. However, there probably was not an established population in the state until recently.

Oklahoma

Cougars were reported in the Wichita Mountains in 1765 by the French explorer Brevel.[32] Early records list cougars as being present along the Alfalfa-Major county line in 1800.[33] Nuttall, in his travels in the region in 1819, remarked that "panthers" were common along the Red and Arkansas Rivers.[34] In 1845 a cougar was spotted leaving a wooded area in a ravine by an expedition near Antelope Buttes in Roger Mills County in northwestern Oklahoma.[35] Woodhouse, as reported in Tyler and Anderson's 1990 account of large mammals in Oklahoma, reported that in 1853 he heard one in a swampy area near his camp in northeast Oklahoma. R. B. Marcy recorded the killing of a cougar in 1852 near the mouth of Cache Creek. In the same year, he also recorded killing one near the head-waters of the Red River of the South and later another one in southwestern Oklahoma near a waterhole along the Red.[36] In their travels through Oklahoma in 1858, Sitgreaves and Woodruff reported seeing cougars but did not specify where.[37] Cougars were also among animals sighted along the Canadian River in McClain County in 1883.[38] There is no official record of the last cougar killed in Oklahoma, but Tyler and Anderson cite a quote that the last cougar killed in the Wichita Mountains was in 1900. From that point it has been assumed that cougars had been extirpated from the state. However, as with Nebraska, persistent sightings, some confirmed, have been reported with the latest as recent as 2011 (see chap. 3).[39]

South Dakota

W. J. Hoffman in 1877 reported that cougars were occasionally captured in oak groves on Oak Creek, Carson County, near the junction of the Grand and Missouri Rivers in South Dakota.[40] Cougars were considered to be numerous in the Black Hills in the 1870s and 1880s.[41] Theodore Roosevelt reported that cougars were also found in the timbered edges of streams in the Bad Lands. Though no cougars had been reported from the Black Hills area since around the 1900s, one was killed in 1931 near the Hardy Ranger Station in Pennington County. They have since then reestablished themselves in the Black Hills (see chap. 3).

Wisconsin

Prairie originally occurred in scattered locations in southern and western Wisconsin within a matrix of oak savanna. Only a small southwestern portion of the state contained original prairie habitat, and no records of cougars are reported from this region. Most reports are from the northern and eastern forested regions with one being killed in 1863 on the headwaters of the Black River in north central Wisconsin. Young and Goldman report a preserved specimen of an animal killed in 1857 near Appleton, which is in the eastern side of the state near the Fox River. This specimen is of interest because it became the type specimen for the cougar subspecies thought to occur in this region, and it came from an area that was historically a prairie-woodland mix. The last animals reported being killed in the state come from more northern, forested counties with one being in 1884 from Ashland County, one in 1903 from Barron County, and the last one in 1905 from Douglas.[42]

WHAT DOES THIS TELL US?

Besides providing information on the demise of cougars in the plains states, what does this death list tell us? Obviously none of the information is new, having been compiled, summarized, and reprinted by others many years ago. I am not presenting anything new when I say that in general, cougars were extirpated from the prairie states from the mid- to late 1800s. These are the data of others and only add to a long litany of species we lost with the "settling" of the prairie. How is this different from many other sad stories of greed, overuse, and insatiability, typical of the frenzy that led to the destruction of the prairie ecosystem? The important difference is in how and where cougars disappeared from the landscape. Unlike the bison who were killed by the millions or the wolves who were hunted and trapped by the thousands, cougars were extirpated by the killing of one here, a couple there. In none of these descriptions was there any mass slaughter, no pictures of hundreds of skins or skulls. Rather than being driven to extinction in a burst of excess killing, cougars slipped quietly into oblivion. In many cases, they disappeared without many people even knowing they had been there.

What does this quiet disappearance of cougars from the prairie landscape tell us about their abundance, their distribution, and maybe even a little about their ecology? First, because of the few sightings and the ease at which they were extirpated, we can conclude that cougars were never very numerous in the vast prairie region. Add this to what we know now about their secretive nature and we can understand why cougars did not predominate in Native American folklore, nor were they prevalent in the experiences of the pioneers venturing out onto and eventually conquering the region. Even many of the earliest explorers, awed by the spectacular herds of bison and the ever-present wolves, probably never saw a cougar. They were indeed the phantoms of the prairie.

But they were there. Why, then, did they not share dominance with the wolves? They were predators and there was abundant prey for them on the grasslands. Why did they not achieve the abundance of wolves or coyotes? Here, the accounts of where people saw, and most often killed, cougars in the prairie region gives us some insight into this ecological riddle. What we notice first in these accounts is that cougars were rarely seen "out on" the grasslands. Most of the references to sighting cougars or where they were killed indicated forested areas or along rivers, which were mostly wooded. Nuttall noted they were common along the Red River in Oklahoma. In North and South Dakota, Nebraska, Missouri, wherever the Missouri River flowed, cougars were sighted along its forested waterway. They were killed in river bends or in forested areas such as eastern Iowa. They were heard in swamps. If a state had a forested region, for example the Black Hills or the forests of Minnesota and Wisconsin, that was where cougars were considered more common, though never mentioned as abundant.

What can we determine from all these observations by so many people over this large area? The obvious conclusion is that cougars in the plains regions were river cats. They lived not on the grassland with the wolves and bison but along the rivers that cut across and intertwined with the vast grassland. Only in the forested anomalies such as the Pine Ridge area in Nebraska or the Black Hills of South Dakota did they seem to venture away from their preferred riparian habitat. This simplified conclusion provides us with the answers to questions as to the abundance, distribution, and role of cougars in the prairie region.

Because they restricted their use to the river habitat, which was a fraction of the total area of the plains, they did not partake in the abundance of prey roaming the fertile grasslands. Thus, they could not build up large populations as did the wolves. Because of their limited numbers and distribution, cougars were probably very susceptible to extirpation. As with wildlife, river banks were attractive to early settlers. Because of their more fertile soil and availability of water, riparian areas were often the first habitats to be degraded by grazing or converted to farmland. The general pattern of cougar loss first started with reductions of their prey base by hunting or habitat loss. This led to declines in cougar numbers, which made these reduced populations further susceptible to extirpation. The actual patterns of loss probably followed the waves of settlers as they proceeded upstream from the East.

Why did cougars not use the abundant prey available to them on the prairies? Why did they instead choose to remain in the river bottoms, restricting their distribution and thus abundance? Would not evolutionary forces favor ancestors that could partake of this abundant feast, freeing future generations from the confines of the river banks? This, however, did not seem to be in the evolutionary cards for the cougars. Instead, the evolutionary forces on the plains evidently favored those who restricted their movements to the rivers. In a land of plenty, something kept cougars from partaking in the main banquet, constrained them to the thin blue/green lines that reached all around but not onto the main grassland table. What could that force or those forces be? The answer to that has to be in their evolutionary history, their ecology, and in exactly what it means to be a cougar. By exploring the behavior and ecology of cougars, we can build a better understanding of how cougars survived in the pre-settlement plains of North America, their role in the plains ecosystem, and, as importantly, what all this means regarding the possible return of cougars to this vast region.

2 Ecology of Prairie Cougars

HAVING SEEN THE HISTORY OF COUGAR POPULATIONS in the Midwest, and what happened to them, we need to see what this has meant for the region. To say that the prairie lands of the Midwest have been altered by human activity in modern times is an understatement. The grassland ecosystem has literally been dismantled piece by piece. Each original piece not of value to the arriving settlers was discarded. With each loss, the once-functioning ecosystem began to unravel. Lands laid bare by the plow and whipped by unrelenting winds strip the soil and pile "snirt" (snow and dirt) in the ditches of northern areas. The draining of millions of acres of marshes and wetlands results in falling water tables and dwindling waterfowl numbers. Straightened rivers and streams rush water draining from the landscape to flood downstream lands. Over-fertilized soil carried by the rivers southward feed the growing dead zone in the Gulf of Mexico. The loss of millions of migrating bison, equivalent to the herds of wildebeest in Africa, impacted nutrient cycling and energy flow patterns of the prairie in ways we are just now realizing. All of these demonstrate an ecosystem out of balance, deteriorating to a shell of its former greatness.

But what about the cougar? What can we say about the role of the cougar and how important it was in the prairie ecosystem? What did we lose with its disappearance from the landscape? What impact did the demise of cougars have on the ecosystem of this region?

We need to have some idea of the original ecological role of the cougar in the plains ecosystem. Was it a minor actor, whose passage from the stage was hardly noticed? Or was it a star with a leading role that was pivotal in the functioning of the prairie ecosystem and whose loss had a significant impact on

29

ecosystem function? Once we know this, we can assess the importance the loss of cougars was to the prairie region and how important it might be to bring them back to the landscape.

The ecological role of any species depends, first, on its distribution and abundance within the various habitats available to it, and second, on what part it actually plays in ecosystem dynamics. As for distribution, cougars originally were the most widely occurring large predator in the Western Hemisphere. Before Europeans arrived, cougars were found as far north as southern Canada and as far south as Patagonia in South America. Within this vast range, they lived in all habitats imaginable: steamy tropical forests, cool temperate forests, rugged mountains, flat coastal plains, dry shrublands, harsh deserts, and prairie grasslands. So cougars are common actors in a wide variety of ecological stages.

And their role on these stages? That role depends on what they do. What they do is live by killing other animals; cougars are predators, a fact that we humans as fellow predators find hard to accept. In there lies the heart of the ancient feud between us and "them." Unlike many predators, though, even the mighty grizzly, cougars are strict meat eaters, no roots or berries for them. In all the prey I found killed by cougars, the one distinguishing characteristic of the site was the stomach meticulously removed and cast aside, as in disgust, from the rest of the cadaver. Bones too were often left behind, stripped clean of the muscle by the cougar's raspy tongue. Cougars indeed are the quintessential carnivore.

Not only are cougars predators, but they are predators of big prey: guanacos to the south and deer and elk to the north. They are also solitary predators, easily killing prey twice their size or more, and they do it quickly. In my tracking of cougars in the snow, commonly in ten to fifteen meters after cougar and deer tracks meet, the drag marks begin as the cougar pulls the lifeless carcass away. Within that brief distance and time, they are able to not wound but kill their quarry. Their efficiency was often admired by the houndsmen who helped me, commonly comparing cougars more favorably to wolves and coyotes whom they despised because these predators seemingly prolonged the agony of their prey, feeding on them before they were dead.

In all areas, this pure flesh diet of large animals indeed puts the cougars at the top of the food chain. Their role, then, is as apex predators, feared by all except jaguars, wolves, and bears. As such, they are a keystone species, defined as having a role whose importance in the functioning of the ecosystem is beyond that implicated by their numbers. They are also an umbrella species, whose habitat requirements embrace those of many others. All of these ecological terms have been used to describe the position and role of cougars in the vast ecosystems of the Americas. In the dynamics of most ecosystems, cougars are not minor actors, not stand-ins, not faces in the crowd. Their role is significant, and many feel that if not the stars of the show, cougars are at least major supporting actors. In many habitats, as other large predators, wolves and grizzlies, were removed by man, the role of cougars has taken on even more importance. Apart from the omnivorous black bear, over most of the western United States, cougars are the only true large carnivore left. As we learn more and more about cougars in these areas, we continue to appreciate even more how vital that role is. Was this also true for cougars on the prairies?

Ironically, in the pre-settlement prairies, where millions of ungulates—bison, elk, pronghorn, and deer—roamed, the cougar at first would appear to have had a limited ecological role. Overshadowed by the ubiquitous wolf and the grizzly bear, cougars were rarely noticed or mentioned by early explorers (see chap. 1). This is because, in this land of ungulate abundance, cougars were probably at their lowest densities of all habitats in which they are found. How could this be? Are they not skilled predators of ungulates and were there not millions of ungulates on the prairies? Why wasn't this immense prey base reflected in cougar numbers?

One explanation for this paradox is that, as solitary hunters, cougars were no match for bison. This excluded them from partaking in this vast quantity of food, which the pack-hunting wolves could exploit. In addition, there were also millions of deer and elk across the plains region, probably more than found in the mountains to the west and the forests to the east. In those areas, cougars easily preyed on these species, but yet, in the prairies, cougars remained left out.

To understand this seemingly ecological contradiction, we need to see where and why cougars were found within the prairie ecosystem. Historical records indicate cougars in the prairie region seemed to be primarily found along river systems (chap. 1). Rarely were they found in the open prairie. Based on this, we can envision cougar distribution in general as being somewhat linear and fingering out across a state's waterways. As today in dryer areas in current cougar range, individuals probably moved up and down the river breaks in their daily movements and when they dispersed to new areas. The river then can be considered the defining element in the life of a prairie cougar. In the plains, "river lion" would have been a more appropriate name than the commonly used "mountain lion." Depending on distance between rivers and intervening habitat, little movement between rivers over the open prairie likely occurred. Thus, the majority of the prairie region was not used by cougars, excluding their access to the bountiful feast that supported large numbers of other prairie predators, including humans.

But why would cougars limit themselves to the waterways in a grass sea of wildlife bounty? The answer lies in the two main goals in the life of any animal: to capture food and to avoid being food. Each day an animal has to try and find an adequate supply of food, a not-so-easy task for herbivores and even harder for predators. Grass will sit and let a grazer eat it. Grazers on the other hand don't particularly like being attacked and killed by predators. And so they will resist their attacker's efforts. How does this reluctance to be a predator's meal explain why cougars were restricted to riverways?

The types of food any species depends on is defined by the evolutionary limitations in their capabilities to acquire and use different sources. We eat what our ancestors ate. We are results, for better or worse, of ancestral decisions made at all stages of the evolutionary journey a species has taken. These decisions made far in our evolutionary past shape our modern form and function. For herbivores these limitations commonly center on such things as digestibility: grass versus forbs or fruit versus seeds. These limitations can sometimes be linked to specializations in how you acquire your food, such as the beak of the hummingbird, or in how you process it, such as the digestive system of the koala. Given your evolutionary limits to the types of food you use, it follows naturally that, if all else is equal, you look for that food where it is most likely to be found. Browsing animals will favor shrubby habitat, grazers more open grassy areas.

For predators, these limitations more frequently involve abilities to capture reluctant prey, or what we call the predator's lethality. The better a predator is at subduing a prey species, the more lethal it is. Animal tissue is relatively uniform in composition and energy value, so how large or how much the animal fights back that you're trying to eat become more important constraints. Speed and stealth also play important roles. In this case, surprisingly simple evolutionary decisions to use claws for traction, as in dogs, or for grasping, as in cats, determine whether you run your prey down or sneak up on them. Evolutionary decisions also determine what kind of prey you seek and where you look for them. If you evolved to run down your prey, you will primarily look for them in open areas that are conducive to running. If you sneak up on them, you look for cover to hide behind.

How the different habitats within a landscape provide or do not provide the necessary elements or opportunities for success produces what I call the *landscape of opportunity* for the predator. We can envision this landscape as something similar to a topographic map where the "peaks" and "valleys" are now not elevation changes but rather represent areas of different hunting success. The peaks are areas where, because of habitat structure, the predator is highly efficient or lethal. Valleys are the areas where habitat combinations reduce the predator's chances for success. If you run your prey down, your lethality will be low in areas with trees and shrubs. For one that sneaks up on its prey, open areas are to be avoided. So in both predator and prey, there is an evolutionary relationship between *what* you eat and *where* you eat it.

As for trying not to be eaten by others, this complicates the process of finding food. As mentioned, prey are, to say the least, reluctant to be eaten by their predators. On the contrary, a predator's life depends on catching and eating these reluctant prey. This creates a two-player system of cat and mouse where prey actively avoid predation and predators attempt to outwit their prey. The result is a wide variety of strategies and counterstrategies by prey and predator, often likened to an evolutionary arms race. Prey grow or develop antipredator defenses and behaviors; predators respond with stronger jaws, longer teeth, or counterbehaviors. Bombardier beetles (Carabidae) develop a foul spray to discourage rodent predators. The beetle's predator, the grasshopper mouse (*Onycomys* sp.), develops the behavior of jamming the bug's rear end into the sand to avoid being sprayed. The list of these fascinating mutual adaptations is long and makes enthralling material for nature movies.

Besides changing body structure, another, simpler way prey can reduce their risk of being eaten is to avoid areas where predators, because of their evolutionary past, are best at catching them. If your predator is highly lethal in the open, stay near the edge of the forest or in the brushy areas. There is growing evidence that prey will do this. The now-classic example is Yellowstone National Park where, after the reintroduction of wolves, we found elk had abandoned the open areas in favor of the forest edge.[1] Therefore, if someone is out to kill you, avoid where they hang out. Stay out of the dangerous parts of town or, if you must go there, be very afraid.

This all produces what I call the *landscape of fear*.[2] This landscape is essentially the prey's view of the predator's landscape of opportunity discussed earlier: a habitat-mediated change in predator lethality as you move over the landscape. This change in predator lethality produces corresponding changes in the level of risk prey face. It is the landscape of fear for the prey: as the risk of being killed changes over the landscape, so must the level of fear an animal should experience. If an animal does not change its fear levels, it will be killed. The peaks of the landscape are where the predator is the most successful or lethal;

these are the areas where prey should be most afraid. The valleys of low hunting success for the predator are the refuges for the prey, the "safe houses" where you can drop your guard.

In our work with elk in Yellowstone, we saw the dramatic effect of these changes in predation risk and the resulting levels of fear in the elk. Because wolves had been reintroduced only in certain locations in the park, there were, for the first few years at least, areas with wolves and others without them. In viewing the behavior of elk in the two areas, the only human comparison we could come up with was two countries, one at peace and the other at war. In the area without wolves, elk continued to behave as they had for decades, and it looked like a scene from a Disney movie. Elk calves would be running and playing or lying peacefully in the warm sun while their nonchalant mothers peacefully grazed with only an occasional look up to check on their wayward offspring. Meanwhile less than fifty miles away, elk calves were found clinging to the sides of their anxious mothers, who found scarce time to graze as they nervously scanned the horizon for wolves. These elk calves, if they survived, grew up without play, without leisurely afternoons in the warm spring sun. They grew up in a landscape of fear.

It is this landscape that prey must face daily in their search for food. Being killed by a predator takes only seconds, while starvation can take months, so ignoring predation risk in the search of food can be a deadly choice. For the sake of an evolutionary future, prey must be afraid to use those habitats where their chances are highest of being killed. Consequently, prey will use a landscape based on what they eat, but most importantly, on what is out to eat you and where they are most likely to do it.

This is the yin and yang that all animals, including most predators, face: you need food but you must also avoid being food. Even top predators like the cougar are not immune and must also live in their own landscape of fear. In a paradoxical twist of big fleas having little fleas, cougars face danger from others bigger, stronger, or more numerous than themselves. How cougars balance these conflicting demands results in where they live on the prairie and their subsequent role in the prairie ecosystem.

Being cats, cougars are stalking predators. Instead of chasing their prey with the hopes of eventually overtaking them, cougars stealthily approach an unwary prey until they are close enough so that with a short burst of speed, they spring onto the animal and subdue it. The first necessity for this hunting approach is cover, something to hide behind. The second is to have ample cover to sneak up on a prey before it spots you. A single tree or bush to hide behind is not enough.

Cougars seek out and use those habitats that provide the right combination of hiding and stalking cover—not too open, not too closed.[3] This in itself could explain the cougars' avoidance of shortgrass areas: there is just no place to hide. If prey can see a predator coming, why bother even trying to catch them, especially if they can outrun you? So, because of the evolutionary strategies of their ancestors, the bounty of the plains was literally out of reach for prairie cougars. A cougar trying to live out on the open prairie would have starved in a land of plenty.

What remains are the river breaks. These serpentine threads of riparian habitat spreading across the plains provide the shrub and tree cover necessary for cougars to make a living. Ironically, the river valleys of the plains become the peaks, or in this case, the ridges, of the landscape of opportunity for the cougar. It is in these stands of shrubs and trees where the cougar comes into its own, the silent stalker, the shadowy presence, the phantom of the prairie. Here its evolutionary traits for stealth change from a shortcoming to an asset in the pursuit of its prey. Because the ridges of opportunity drop off quickly as trees and shrubs diminish with distance from the river, the cougar has no incentive to venture much beyond its brushy limits.

Although we can see that the shortgrass areas of the prairie would not provide adequate hunting habitat for cougars, areas of tallgrass prairies would seem to be more than adequate. Early accounts of grasses as high as a horse's belly in these areas could easily hide a cougar and allow it to stalk its prey, in a similar manner to the African lion, which uses tallgrass to its advantage. Would we expect higher abundance of cougars out on the plains in these tallgrass areas? We might indeed if it weren't for the second factor affecting where cougars might hunt: wolves.

It has been well documented that wolves dominate cougars, often chasing them away from kills they have made and harassing them in general: the old dog-and-cat thing. In fact, many feel it is because of wolves that modern-day hunters with one or two well-trained hounds can chase a cougar twice their weight up a tree. Whether that is true or not, wolves do provide a threat to cougars, one that they can coexist with if they can escape up a tree or into rocky terrain where wolves cannot follow. Superimposed on the landscape of opportunity for cougars is their landscape of fear in the presence of wolves. Areas where there are trees to climb and bushes to hide in provide security for cougars against harassing wolves; open grass areas, short or tall, do not. Caught out on the prairie by a pack of wolves who can easily outrun you is almost certain death or at least serious injury. This makes the open prairie a dangerous place to be if you're a cougar, an enormous peak in their landscape of fear. Fortunately for cougars, areas where they are successful in hunting also provide protection from their potential predator, a rare case where feeding opportunities and safety coincide. So for food and safety, the cougar becomes the river lion of the prairies.

For these combined reasons, it is safe to surmise that cougars were riparian predators in the great grasslands of America. It is there that they performed their role as a top predator. But because of this limited area of efficiency, and the limited amount of riparian habitat, what was that role and how important was it to the prairie ecosystem? Because they did not use the open grasslands, they did not, and probably could not, hunt the large bison herds like the wolves and humans did. Deer and elk out on the grasslands would also seem immune to and protected from the cougar's fangs and claws. Because of this, was their ecological role a minor one, marginalized because of their limited distribution, abundance, and effectiveness? Did cougars on the plains play an inconsequential part in a drama with a cast list of millions? Were they just a face in the crowd? Was their role so minimal that the loss of cougars was of insignificant ecological consequences? Recent studies in the return or presence of top predators in ecosystems indicate the answer to all those questions is a resounding no.

We had originally predicted that the return of wolves to Yellowstone National Park would have significant ecological impacts that would cascade through the ecosystems.[4] We also argued that these impacts would not be solely because wolves were killing their prey but more because they were scaring them. In fact, given that predators in general only kill 20 percent of the prey they attack, they scare a lot more prey than they kill. Having escaped near-death experiences approximately 80 percent of the time ought to be excellent motivation for prey to learn where they should be afraid and where they have a better chance to survive an attack. Because of all this, we proposed that wolves reestablished the landscape of fear for their main prey, the elk (*Cervus elaphus*).[5] This fear of wolves would prevent elk from using the whole park as their personal grazing area as they had for decades before the reintroduction of wolves.

Indeed, this is what we found in the Lamar Valley of the park. As we collected data on how elk were using the valley in the presence of wolves, the shift in use became clear and attested to the power of fear. When we moved from the forest edge out into the now-risky open grassland, the fresh, dark-brown deposits of elk "berries" gradually decreased and were replaced with older, whiter ones excreted in previous years, before wolves. It struck us that the howls of the nearby wolves were drifting over what used to be a center of elk activity. Female elk in the hundreds would peacefully graze on the succulent grasses and forbs while their calves romped nearby. Now the area was deserted, like a ghost town of the old West; only what the elk left behind, in this case their dried feces, attested to these better times. As in those deserted towns, nature began to repair the damage.

With the elks' restriction to safer areas, the valleys in the landscape of opportunity for wolves, these open risky areas received a long-needed reprieve from the annual mower-like grazing pressure they had received over the forty years of the wolf's absence. We predicted that relief from the constant cropping would lead to the recovery of trees, bushes, and forbs in these areas. Terminal buds of tree seedlings could survive and sprout to greater heights with each passing year. Shrubs could stretch out their multitude of branches in all directions without being trimmed back to dense, stunted, low-growing domes. Myriads of flowers and forbs could sprout, grow, and seed before being clipped back to ground level. This regrowth of vegetation would, in turn, benefit a multitude of other wildlife species.

Though meeting initial disbelief that fear alone could produce such dramatic effects, subsequent work by others verified our predictions.[6] The effects of wolves scaring elk were found to cascade through the ecosystem to levels we could not have predicted. The changes seen have been so remarkable that Yellowstone has become the textbook example of how the return of a top predator, mainly through fear, can help recuperate an ecosystem.

Since these findings in Yellowstone, research by Bill Ripple and Robert Beschta has demonstrated similar effects produced by cougars in other National Parks.[7] In this case, the landscape of fear was of mule deer (*Odocoileus hemionus*), and the habitat was streambanks. Because cougars can successfully hunt deer in the riparian habitat, deer are reluctant to use these areas, sparing the vegetation from their insatiable appetites. Remove the cougar and ecosystem function becomes impaired. Trees grow old without hopes of replacement; diversity of native forbs and grasses is reduced as they are replaced by unpalatable, often foreign, species. Apply these findings to the rivers and streams of the Great Plains, and the role of the cougar in these systems takes on a paramount ecological importance.[8]

Rivers, especially in dry grassland areas, often provide the only source of open water in an otherwise arid landscape. Riparian habitats, because of abundant moisture, are bands of highly productive and nutritious shrubs and forbs running through the seas of low-nutrition, hard-to-digest grasses. These areas would, under normal circumstances, be magnets to the millions of grazers and browsers living on the prairie. In addition, such wooded areas may provide protection against wolves inside riparian areas but would be susceptible to overgrazing/ browsing, as they are today with cows. Given the high densities of and damage done by deer in riparian areas without cougars, we could envision such overuse occurring in pre-settlement times, if not for the presence of the cougar.[9]

Because cougars prowled the streambanks, deer and elk probably did not spend much time in these habitats. This fear-induced reluctance probably spared seedling willows and aspen, and allowed continued replacement of riparian forests; the maintenance of these trees preserved the integrity of the stream environment, providing shade, deep pools, and an influx of organic material; and willows and aspen provided food and shelter for semi-aquatic species such as beavers, muskrats, and mink.

Based on what we know today, we can assume the main ecological role of the river lions was as the guardians of the integrity of riparian habitat. Just as the wolves were the shepherds that kept the bison moving and prevented overgrazing of the grasslands, cougars probably protected the riparian habitat and thus the health of the river and stream ecosystems. Though limited in distribution and overall numbers, cougars were no minor actors in the great prairie performance. Their role may not have been center stage, out in the open grassland, but it was significant and vital to the prairie ecosystem. Without their presence in the riparian habitats, the myriad plant and animal species in and out of the water would have suffered as they were trampled by millions of unabated and unafraid feeding ungulates. Trout and other species of aquatic wildlife, needing the shade and nutrients trees and bushes provided, would have found the river a hostile place to live. Beavers, due to the lack of regenerating aspens and willows, would not have been in the abundant numbers found by early trappers. And even the ungulates themselves that relied on the riparian habitat, through their destruction of it, would have suffered. Rivers are often referred to as the lifelines of the prairie ecosystem; those lifelines in the past probably owed their survival to the cougars.

Though ecologically important regardless of their abundance, it would still be of interest to ask just what might have been the cougar's population numbers across this broad area. This may seem to be a formidable task, especially given their restriction to the riparian habitats of the region. But it is specifically because of this restriction that our task becomes easier. One of the advantages of viewing the landscape of opportunity is that we can not only identify where successful hunting habitat is for cougars but we can also quantify it. Once we quantify it, we can then estimate how many cougars might occupy it. In general, we would predict that the larger percent the landscape of opportunity consists of high-quality hunting habitat, the larger the predator population will be, up to a point. If there is too much successful habitat, prey will have a hard time surviving and so will the predator. In this case, too much of a good thing is not so good for a predator.

In the prairie areas, however, the opposite is the case. The large blocks of open grasslands provided vast amounts of safe habitat where prey could escape the stalking cougar. Being restricted to the high-quality hunting habitat of the river systems, the number of cougars then becomes closely related to the length and width of available riparian habitat. If we can estimate how much of that habitat existed, we can begin to estimate the number of cougars that might have roamed these areas.

Based on original stream and river distributions across the Great Plains, I have reconstructed where cougars were likely found in pre-settlement times. The reconstruction is just a rough estimate since little is known of original stream-bank vegetation in those times. It is hard to use present-day conditions because so much of the original riparian vegetation has been altered. Out of necessity I make a simplifying assumption that on average, riparian habitat will extend from the river or stream's edge to one hundred meters out on each side of the river, giving us a riparian area two hundred meters wide along all rivers and streams in the prairie region. Surely there were stretches of river without woody vegetation, and other stretches, especially in the larger rivers, where riparian vegetation probably extended much farther. But the goal of this exercise is not to identify which rivers and streams might have had more riparian habitat and thus more cougars. The goal here is to get an estimate, as rough as it may seem, of what might be a reasonable total number of cougars across the whole plains area. That is, were there thousands or hundreds of thousands of them?

If we assume the areas of wider riparian habitat balance out the areas without woody vegetation, then an average of two hundred meters might be reasonable. We can increase or decrease this estimate to provide a "sensitivity analysis" designed to give us probable maximum and minimum numbers of cougars; this will determine if we are talking about a lot of cougars or maybe not so many. Also, I did not initially include the obvious islands of cougar habitat such as the Black Hills of South Dakota or other forested patches on the plains. Here our primary interest is to estimate how many cougars might have been found along the rivers "out in the plains."

Based on estimations from satellite images (fig. 5), rivers and streams criss-crossed much of the midwestern prairie habitat. Thanks to the magic of satellite images and Geographical Information Systems (GIS) software, we can calculate

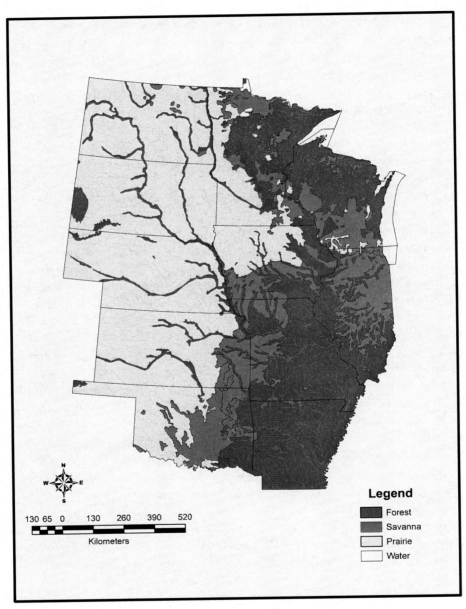

Legend
- Forest
- Savanna
- Prairie
- Water

130 65 0   130   260   390   520

Kilometers

FIGURE 5. Original vegetation across the prairie states; note the network of rivers that crisscrossed the plains, allowing access to these regions to cougars. (Map based on work from Kuchler 1964 and digitized by the U.S. EPA.)

the length of the streams and rivers in question. If we consider just those permanent flowing waterways, a rough estimate gives us approximately 217,000 kilometers of rivers and streams in the prairie region. If we assume an average riparian width of 200 meters (100 on each side), we come up with 43,500 square kilometers of possible cougar habitat along the rivers in the Midwest. Cougar densities (number per 100 square kilometers) vary widely from less than one to more than three.[10] The maximum density of cougars an area can support is related to the amount of hunting habitat available.[11] If we assume the estimate of riparian vegetation represents ideal hunting habitat, we can use the higher density estimates of cougars, three per 100 square kilometers. At this density, over all of the prairie lands of the Midwest, there would have been approximately 1,300 cougars. If we double the width of the riparian corridors to 400 meters, we essentially double our estimate of cougars to 2,600. Even at an impossible total riparian corridor width of 1,000 meters, the estimate is only around 6,500 cougars.

We will never know which of these estimates is closer to actual pre-settlement abundance of cougars, but it does demonstrate that in all cases the number was not very high: 1,300 or 2,600 cougars may sound like a lot, but these two estimates are for a 3.6 million square kilometer area encompassing most of nine to ten states. These numbers are less than reported for within a single state in current cougar range to the west. For example, in Idaho, the estimated population is 2,000 to 3,000 animals in an area approximately one-fifth the size (216,600 square kilometers). Thus, anyway you figure it, cougars were not plentiful over the plains region and now we know why—because of the limited availability of successful hunting habitat.

Because of local habitat anomalies, there were probably some areas on the plains that supported more cougars. The Black Hills of South Dakota currently has a population in excess of 130 adult individuals; historically, though, the river systems of that state probably supported only another 200 cougars, primarily along the Missouri River. Also recently, cougars have established themselves in the Badlands of North Dakota. Again, even with these islands of cougar habitat, total numbers were low and probably never exceeded 3,000 for the total plains region.

Given that most cougars in the plains lived in riparian areas, how did these cougars "fit" into this vast linear, winding network of riparian habitat? Did they move freely up and down these aquatic corridors or did they space themselves out in some form of social hierarchy? If there was some sort of social order, how did this affect the cougars' ecological role and success in the plains ecosystem? To answer those questions, we first need to know something about home-range dynamics of cougars.

Originally, it was believed cougars were fiercely territorial, fending off all intruders and insuring their sole use of an area. The home range, a place to be familiar with, was the same as the territory, a place to defend. For male cougars this thesis has more or less stood up to current scientific rigor. Males do seem to stake out a territory, which they defend by scent marking and, judging from the scars on their faces, numerous physical battles. However, how successful they are at doing this is still up to debate. Most recent studies of male home ranges indicate little overlap of core areas, but borders are poorly defined and overlap extensively. This makes the sizes and shapes of these territories dynamic and very flexible. At any given time, existing territories depend on how strong you are, how weak is your neighbor, how valuable are the resources you have, how valuable are your neighbor's resources.

If you think this sounds like medieval Europe, you're correct: the dynamics are similar. Sizes and shapes of "kingdoms" depended on the personalities of individual rulers, the resources they had available, and how defendable the landscape was, all of which resulted in changing border lines and disputed or overlapping areas that resulted in battles and battle scars. Male cougars, as like many human despots, rarely get along with their neighbors and are constantly either defending their territory or trying to expand it. Consequently, the territories of male cougars can be envisioned as vacillating kingdoms with shifting borders, boundary skirmishes, outright invasions, and coups paralleling the dynamics and intrigues so characteristic of human history. As with humans, it appears that male cougars consider the territory a piece of land worth fighting for.

For human territories, the resources fought over were often land-based: agricultural land, minerals, water, space, and the like, resources we need (or think we need) to survive. For male cougars, what would be so valuable as to risk injury defending or trying to take over a "piece of land"? Surely it could not just be to have a place to hunt; there is usually plenty of deer and other prey to support even randomly wandering males. Why then, would male cougars put so much effort into and take so much risk in maintaining their territories?

Modern population genetics can provide us with an answer. We know that for most species, a male's genetic contribution to future generations is in no way assured. In contrast, most females surviving to reproductive age will have a chance at passing on their genes. Few females do not mate and thus their challenge for genetic survival becomes the survival of themselves and their offspring. For males, study after study has demonstrated that in most species, only a small percent of the males reaching reproductive age actually mate successfully. All males are willing, but few are selected because it is the females' choice: they decide with whom they care to pair up their genes. They must choose well because, unlike the male, their chance at genetic survival is limited to the one to three litters of kittens they will have in their lifetime. Consequently, for sometimes obvious and other times not-so-obvious criteria, the majority of males are excluded from mating opportunities. To be the lucky few, males then must play the game of meeting the criteria of the females.

This raises the age-old questions: What is it that females want? What are they looking for in a good male? For female cougars, the one important criterion seems to be the territory. Few male cougars have opportunities to mate if they do not have a territory. For male cougars, then, the territory becomes a must-have if you're going to pass on your genes. This means the "resource" that makes a particular territory worth defending is access to females. Since this is the ultimate resource for passing on your genes, we can understand why males will risk their lives to get this kind of territory and then defend it at a high cost. Without it, even if they survive, they are genetically dead.

An added benefit to the male for maintaining his territory is that his presence also aides the survival of the offspring he sires. Infanticide by male cougars is well documented; there are indications that they primarily kill kittens sired by other males. In a manner similar to African lions, the presence of the male reduces the chance an intruding male may find and kill the resident male's kittens. In fact, this may be the main reason why females select males with territories. For the male however, we can ask whether this is just a fortuitous side effect from defending his mating rights or an added conscious effort to protect his genetic investment beyond the insemination stage. Unfortunately, we know too little about the social interactions of this solitary species to say for sure. Even in African lions, judging whether males' efforts to ward off other intruding males are just to preserve breeding rights or to protect their cubs is a call fraught with anthropomorphism. All interesting questions for future cougar and lion biologists to investigate.

Out of necessity, then, a male cougar's territory/home range has to overlap with the home ranges of female cougars. The question then becomes: How many females is enough? Obviously the territory has to provide access to enough females to make it genetically worth a male's time to defend it. The gestation time for cougar kittens is three months, and the kittens stay with their mothers for up to sixteen months, so it takes a female cougar almost two years to successfully raise a litter of two to three kittens. A male, then, gets to pass on his genes to a particular female only once every two years. Because each year there are younger and progressively stronger males trying to take over an aging male's territory, a male is lucky if he can hold on to a territory an average of four years. Having to defend that territory for the opportunity to mate only twice over that time does not seem much of a genetic return on your investment.

Increasing the number of female home ranges your territory overlaps increases your genetic payback for a minimal expansion in your investment. We really don't know how many females makes it worth it for the male, but most home-range studies find extensive overlap of male territories with usually three to four females. The sizes and overlap of home ranges of females then sets the size of the territory a male should defend, or try to. If females have exclusive territories, a male will have to defend a much larger area than if female areas overlap extensively. This impacts the resulting spatial arrangement of male territorial cougars over the landscape. In the case of the prairie region, this will determine how many males and the way they are distributed along the river corridors.

Now we turn our attention to the home range and territorial behavior of the females. If we could estimate the size of an area female cougars need and whether they defend this area, then we can determine how females "fit" into the riparian habitat. We then can overlay the territories of the males. Here, however, the picture becomes even messier. Where we find male territories overlapping on the borders, we find females seem to lack much territorial behavior at all. Study after study of female home ranges show extensive overlap existing among several females. The main reason for this difference from males is that females don't need to secure and defend an area to find a mate. As is quite universal in the animal world, the males come looking for them. Excluding or not excluding other females will not enhance or reduce a female's opportunity to mate.

A female's major concern, then, is how to successfully raise her offspring. For cougars, the male does not help, except for the indirect protection he affords from other males. For females the important consideration becomes how an area provides prey resources and a safe place to raise their young. What are the criteria for this selection and just how big of an area is needed to do this? First of all, it is obvious by the overlaps in home ranges that sharing a large part of that area, whatever size it is, with other females does not seem to be a major concern. This, in a sense, would seem counterintuitive from a hunting standpoint.

How many of us hunters or fishermen have fumed and complained when we see others using the same area? I know of hunters who will abruptly leave an area upon seeing another hunter. We know (instinctively?) the more hunters there are, the less there is to go around and warier the prey will be. So why is there the overlap in home ranges of female cougars? Wouldn't a defended hunting territory provide increased hunting success for a female?

I think the answer is that it is not just other hunters but who those hunters are that makes a difference and helps explain this overlapping use by females. Though we may not like it when strangers intrude on our hunting or fishing experience, most times friends and family are welcome. In a similar manner, recent research has demonstrated that most female cougars who are sharing an area are related. They are either a mother and her grown daughters, or sisters, or at times cousins or aunts and nieces. As we recognize in human societies, sharing among family is okay; somehow this is not the dreaded socialism. What appears to be happening is a tradeoff between exclusive hunting opportunities and the protection provided to offspring by being within the territory of a male. However, you can reduce the costs of sharing hunting opportunities by sharing the area with related individuals. This raises the interesting question of whether females would exhibit territorial behavior against nonrelated females. Or will territorial behaviors flare if too many female relatives are in an area? Again, these are questions that await additional research on cougar social behavior.

Given that several females will use an area, how do they share it? Do they each move around randomly, possibly crossing paths, or worse, interrupt each other's hunting attempts? Probably the best way of viewing this overlap by females is by noting that rather than just mutually occupying the same area, they are cooperatively sharing the use of it as a loose family group, similar to an African lion pride where females are related and share the resources of the area. There is no evidence, though, that female cougars actually cooperate in hunting or sharing captured prey, as lions would do. In fact, simultaneous relocations demonstrate a certain amount of temporal separation, with females rarely being in close proximity of each other. What it appears to be is just a higher tolerance of other related females in the area and, though currently undocumented, may involve some form of olfactory or vocal communication coordinating their movements.

At times mutual use seems to involve more than just keeping out of each other's ways. I observed one female who was sharing her home range with one of her daughters. When the daughter gave birth to her first litter, the mother, now grandmother, shifted her center of use off to one side of their mutually used area, as if providing more exclusive use of part of the home range to her daughter. Whether that is being too anthropomorphic, I am not sure. I suspect that with the use of new GPS technology where we can keep track of hourly movements of collared cats over months, we will discover new and fascinating details of social behavior of this seemingly solitary predator.

We now know that females share an area with their relatives, but it is of interest to ask why they would want to use the same area over and over again. Why not keep moving over the landscape, exploring new and possibly better areas to hunt prey that may not be aware of your presence and so are easier to catch? The answer to this question strikes at the very basic reason why animals have home ranges. Ever since we could identify individuals, we have observed that they indeed use the same area over and over, often throughout their whole lives. I kept track of one female cougar for eleven years, and she never left the area where she was born. Over this time she shared this area with her mother and subsequently her daughters and granddaughters.

In 1943 William Burt was the first to officially call this the "home range" and describe it as an area in which an animal travels in its "normal activities of food gathering, mating, and caring for young." In doing so, an animal becomes familiar with this area, which is key to this phenomenon. It seems any possible advantage of wandering for more abundant and more catchable prey is overwhelmed by the advantage of knowing details about an area for those normal life activities. For female cougars we can quickly see one advantage of familiarity with their area of use: they would know where good den sites might be, possibly the same one they were born in.

Probably an even more important advantage to familiarity with an area is that individuals will also learn the better places to hunt. Just as we return to our favorite hunting or fishing spot, cougars, based on their past successes and failures, probably build a mental hunting map in their heads. This reduces wasted time visiting areas where prey can't be found or are difficult to capture. This can be helpful for a female anytime but may become critical when she has kittens to feed.

During the thirteen to sixteen months that kittens stay with their mothers, mom does all the hunting. Even at the beginning of this dependency, the female has to capture more food than she would normally need. This added demand borders on the ridiculous toward the last few months when she has to essentially capture enough food to feed three to four adult-size cougar "kittens." Over the period of dependency of kittens, a female needs to capture more than twice as many deer to feed her growing family.[12] Familiarity with her hunting grounds probably helps her accomplish this Herculean task.

How does she do it? How does she become familiar with her hunting grounds? She probably does not learn it from her mother. From tracking cougars in the snow, I have come to learn that the pattern is for the female to leave kittens, even ninety-pound, thirteen-month-old ones, at a kill she has made. The female then goes off and makes a subsequent kill and returns to guide the kittens to this new food source. By repeating this process, kittens rarely participate in or even observe where and how mom gets dinner.

One exception to this happened when I was following the tracks of a female and two kittens in the snow. All of a sudden, the kitten tracks stopped and the mother's continued on. About one hundred meters farther on, I found out why. The mother had spotted a deer in the distance, left the kittens where they were, and began her stalk. I can imagine the kittens on their toes and out-stretched necks watching as mom slowly crept through the snow toward the deer, hoping that mom would be successful. She did succeed that time, the kittens fed well, and learned a little about where and how to hunt. However, this was probably more the exception than the rule.

After leaving their mothers, the more usual process is one of hits and misses, making mistakes, and savoring those first successes. Another observation involved a female I had tracked since she was a kitten. Shortly after she separated from her mother, I came across what was probably the very first deer she had killed on her own. This was an exciting event for me because I had evidence that she could indeed survive on her own. However, I imagined how exciting it must have been for her. And if they would do such things, she would have performed a "touchdown dance" that would have put most football players to shame. As exciting as it must have been, that was just the start of her training.

Genetics gives cougars the basic tools to kill a deer, but practice gives them the experience of how, when, and, most importantly, where to use these tools successfully. All predators, big and small, are formidable threats to their prey, under the right conditions. This is to say that all predators have a weak point, a limit to their effectiveness. If they did not, they would quickly eat themselves out of house and home. It is this weakness in their hunting capability that produces the landscape of opportunity discussed earlier. No predator can do it all, can be effective in all habitats, against all prey types. The newly independent young cougars need to learn that. By near and not-so-near misses they learn that deer are not as easy to catch if they can see you first. They come to learn that hunting in open areas, even if there may be a lot of deer, is a waste of time.[13] Through their successes they learn that edge habitat, at least in forested areas, is where they will catch a meal. They learn what constitutes successful hunting habitat.

As importantly, they also learn where that successful hunting habitat is. Based on past successful/failed attempts, they learn to avoid low success areas and return to the higher success ones. They can learn the movements of the deer, where they go in the winter, where they would most likely be. They learn their landscape of opportunity. On doing so, they gain an efficiency in hunting that would not be possible by wandering over the landscape. So the biggest advantage of staying in one area for a female cougar is learning her hunting grounds, her home range, and she learns how to use it. This is the value of the home range to a predator like a cougar.

If we apply this information to past cougar populations in the plains region, we no longer visualize a structureless population of cougars wandering up and down the river corridors. Rather, we can envision a certain order in the distribution of female cougars along the riparian habitat available to them. We can assume that female cougars in the river corridors distributed themselves relative to adequacy of the habitat to provide hunting opportunities, as did their cousins in more expansive habitats. As we now know, home ranges of related females probably overlapped in a temporally fluid manner. Some stretches of habitat, because of their makeup and structure, might provide better conditions while others may not. Worked out through trial and error, and honed by experience, the thousands of miles of riparian habitat would be carved up into overlapping home ranges that provided female cougars the best chance for them to provide food for themselves and their young.

Males then, would superimpose themselves on the resulting pattern of female home ranges to produce territories that were reproductively advantageous. Some areas, because of habitat, might be "richer" with more overlapping female home ranges. Consequently, there could be a dominance hierarchy with the "best" territories going to the biggest, strongest males. More subordinate males would have to settle for territories with fewer females present. Finally, overlaid on all this would be the annual dispersers, mainly young males, who would indeed wander long distances up or down the streams looking for territories of their own. They would either be successful or be forced out into the prairie where they would disappear from the population, victims of starvation or attacks by wolves. The only difference between riparian and non-riparian home range and territory dynamics is that the riparian ones would be long and narrow, confined by the vast expanses of low opportunity and dangerous grasslands.

The number of these home ranges and, consequently, the number of cougars the riparian habitats of the prairies could support depends then on how large the female home ranges would be. That leads us to critical question: How big is a female cougar's home range? Initially, it was thought that the answer was in the number of prey there were. The higher prey densities, the smaller of an area a cougar needed to catch sufficient food. Even though this belief is still commonly held, with the recognition of the landscape of opportunity we are beginning to realize that it is not so much the abundance of prey but rather the catchability of that prey within the landscape. Even if prey are numerous, if there are not the right conditions to catch them, effectively there are few prey available. Because not all habitat is equal relative to hunting success, it would make sense that the more high quality hunting habitat in an area would affect success rates and ultimately the total area needed for the home range. If a cougar needs to have a certain amount of quality hunting habitat to be successful enough, than her home range area would have to contain the minimal amount of that habitat. Based on this, we would predict that females would need larger home ranges in areas with a lower percentage of quality hunting habitat over the landscape, effectively needing more area to provide the minimum amount of hunting habitat. So the percentage of quality hunting habitat on the landscape, more than just the abundance of deer, will determine home-range sizes for cougars.

For cougars in forested areas, we now know that quality habitat is edge habitat. So the prediction becomes more specific: the more edge habitat an area has, the smaller of an area a female needs for her home range. We tested this idea in Idaho and indeed found support for it.[14] We even found support for the idea that in defining their home-range boundaries, females appear to configure their home ranges to try and maximize the percentage of edge within them. What we did not find is a difference in the amount of edge habitat within home ranges; on average, females had around ten square kilometers of edge habitat, regardless of home-range size. This supports the idea that it is not the number of prey available but the amount of habitat where you can successfully hunt prey that is important when a female cougar defines the limits of her home range.

But why would they need a certain amount of high-quality habitat to be successful? This has to do with the deadly games of cat and mouse that cougars and deer play, and it returns us to the landscapes of fear and opportunity. We know that prey should be fearful of their predators, but they also have to balance this fear with their need to look for and eat food. This puts the prey in a delicate balance of being observant enough to detect predators but also devoting sufficient time to eating. Unfortunately, it is difficult to do both at the same time. In finding this balance, a prey has several approaches it can use. The first is to ignore the predation risk, eat as much as it can, and hope the predator kills someone else. Obviously, this strategy only works if everyone uses it. If you are the only one taking the risks while your friends remain safe, you will be the one to be killed.

Though this seems foolhardy, it appears that males of some species may actually use this strategy. In our study of elk in Yellowstone Park under predation risk by wolves, male elk did not increase their vigilance in the presence of the wolves.[15] We hypothesized that because mating success in male elk is dependent on body size, males could not afford to decrease eating to watch for predators. If a male elk reduced his food intake to be more vigilant against predators, he might not grow as big but will increase his chances of living. However, he is not assured that other males are doing the same thing. The other males may have a greater chance of being killed, but if they survive, they would be bigger and stronger and most likely to win access to the females. As long as some males take this risk by not being vigilant and survive, no male can afford to be vigilant and expect to mate. This locks all male elk into a sinister contest of chicken where each is betting on the misfortune of their comrades. To play the game increases your chance of being killed, but to not play the game insures your genetic death.

For females, this type of game is genetic suicide. Females are not in competition with each other for mating opportunities and so need not worry that they are smaller than their neighbors. Females just need to eat enough to be healthy to give birth to their young and stay alive to do so. Extra food buys them no great advantage. For them, predation risk becomes paramount. To ignore that is to put themselves and their young in danger. Not surprisingly, females should take a second strategy where they balance predation risk and food needs. This strategy involves knowing when to be more afraid and when you can lower your guard to eat more. It involves having a finely tuned sense of whether you're in immediate danger from your predator.

Because a predator can't be in all places at all times, the actual risk a prey faces fluctuates, depending if the predator is near or off somewhere else hunting others. In the absence of danger, it is a waste of time to be on high alert, time that could be spent feeding. Smart prey will be attuned to whether a predator is near and adjust their vigilance levels accordingly, providing more time to eat when indications are that the predator is away. When the predator returns, smart prey begin to be more wary again. How do they know the predator is back? Subtle cues, such as fresh odors or unusual bird activity, and not-so-subtle cues, such as being attacked or noticing that the deer you foraged with the night before is missing, make individuals aware. This information then passes on to others, and eventually everyone knows the predator is back. Vigilance levels shoot up and stay high until the predator leaves the area again and the prey can relax, a little.

How does this benefit a female cougar and how is it related to the size of the home range it needs? Fortunately for the predator in this game, the knowledge of whether a predator is near or away is not instantaneously gained by the prey. It takes time for this information to pass around to all prey individuals living in the area. This means the prey will continue to be relaxed for a short time after the predator has returned. It is during this brief period of time when vigilance is still low that the cougar gets its chance. Just as opening day of a season is the best day for success by human hunters, so it is when a cougar returns to a part of its home range after an absence. Deer have dropped their guard; they are using those more dangerous places, spending more time eating. In the absence of immediate danger, they become lax and more vulnerable. If this sounds like human behavior, the process is the same. We are always the most attentive drivers just after a near accident, not before. And the more time having passed since that near accident, the less attentive we become again. After the first attack or if the cougar is sighted, the guardedness of the deer shoots up, and the cougar's chances of catching prey decline. If the cougar is successful, it gains a meal; if not, with all the deer on alert, it is a waste of effort staying around and eventually it becomes time to move on.

Meanwhile, in the part of the home range where the cougar just came from, deer are beginning to relax, and so it goes: as a cougar moves about its home range, wariness of the deer rises and falls. A smart predator will take advantage of this and manage fear levels to its benefit.[16] Human hunters will often let an area go "fallow," give it a rest. This allows time for the prey to think that the danger has passed, lower its defenses, and upon their return, the hunter's chances of success are higher. Smart predators should do the same. They should move about their home range so as to allow each part to rest sufficiently. In their absence, prey vigilance declines to the point that upon the cougars' return, they can make a kill.

Do cougars do this? Are they smart predators? The fact that their lives depend on it and that, as a species, they have been around for thousands of years indicate that they are doing more than just randomly wandering around their home range. The twelve years of data that I collected on cougars and

deer supports this idea.[17] I found that female cougars showed distinct patterns of movement and habitat selection consistent with trying to maximize
their chances of success. Because it would take longer for all the deer in a large
patch of forest to know about the arrival of a cougar, a cougar would have more
time before the word gets out. Thus, cougars should and do visit larger forest
patches, containing more edge habitat, more often than smaller ones, and
should and do stay longer in the larger patches. This is also reflected in their
success rate, which is higher for the larger patches. It appears, then, that cougars
are at least conscious of or may actually be manipulating fear levels in deer.

To manipulate fear levels in this manner requires an area with adequate
amounts of hunting habitat; otherwise you visit each area of the home range
too often. Frequent use doesn't provide enough time for deer to lower their vigilance levels between visits, and you are faced with alert and hard-to-catch prey
upon your return. Home-range sizes for female cougars, then, likely reflect the
need for adequate hunting habitat to manage the fear and thus, vulnerability of
the prey they depend on.

The more we learn about the hunting habits of predators, the more we
can see that the game keepers of the old world were cognizant of these habits.
The game keepers had immense knowledge of the wildlife under their care
and thus were able to manage the hunting pressure to insure hunting success
by their employers. Or more likely, being predators themselves, game keepers
patterned their actions on what has worked in the context of evolution for their
four-footed hunting brethren. In either case, lives and livelihoods depend on
managing fear and having an area large enough in which to do it. The sizes
of female cougars' home ranges and of hunting preserves represent that area
and are dictated by the amount of hunting habitat available. The lower the
percentage of hunting habitat, the larger the area needed.

How can we use this relationship of hunting habitat to home-range size to
estimate the number of female cougars that might have existed in the riparian
areas of the plains? To start, regarding our riparian landscapes of opportunity,
one can surmise that there were two basic "edge" areas running throughout the
length of the waterway: the river's edge and the other edge as the forest broke
away to the sea of grass. Although cougars can swim and probably were able
to kill ungulates as prey crossed rivers or drank from them, the water's edge
habitat is probably of minor importance. It is along the water's edge where
shrubs are often the thickest, hindering a cougar's view and maneuverability. It
is also hard to subdue a deer in the water, especially deep, fast-moving water. In
contrast, where the forest breaks to the grassland, trees and shrubs gradually

give way to open grasslands. The edge is often not as abrupt, providing cougars with a view, but it also has shrubs for camouflage in order to stalk prey as they come off the prairie or leave the river bank. The grassland side of the riparian zone then provides excellent stalking habitat for cougars. Add to this the large reservoir of prey supported by the grasslands that has to rely on the rivers and streams for water, and you have the equivalent of the African waterhole, extending for thousands of miles across the plains.

Given the relatively narrow width of riparian areas and the length of the rivers, the riparian habitat in the plains areas probably provided the highest ratio of edge to forest anywhere in cougar range. If we assume, as we did in Idaho, that edge habitat extends approximately 25 meters out into the open and 15 meters into the trees/shrubs, it consists of a band 40 meters wide.[18] These distances are based on the visibility a cougar needs to see its prey and on sufficient cover to stalk and successfully attack an animal once spotted.

If we assume an average width (on one side of the river) of 200 meters of riparian habitat, then the total width of usable habitat is 225 meters (200 m + 25 m). For every kilometer (1,000 m) there are 225,000 square meters of total habitat of which 40,000 square meters is edge. This converts to an 18 percent edge within this band of habitat 1 kilometer long by 225 meters wide. Based on the regression equation we developed for Idaho, a female cougar would need a home range of approximately 70 square kilometers of riparian habitat. If we assume a 60 percent overlap of female home ranges, three females would occupy an area of approximately 150–200 square kilometers of riparian habitat.

Given the earlier estimate (43,500 square kilometers) of available riparian habitat, a rough estimate of the average number of female cougars living in the plains area would be 650–1,160. If we divide this number by three, the resident male population would be 220–390. If we assume 21 percent of the total population are transients without home range or territory, then the number of these transients is 230–400 (number of females + males divided by 0.79 minus the number of females + males = the number of transients).[19] Adding them all together would give us an estimate of 1,100–2,000 for the total adult population, minus dependent kittens, inhabiting the waterways of the plains region. This agrees quite well with the estimates derived earlier, which were based only on gross densities.

Besides corroborating my previous mathematical efforts, what do all these mathematical gymnastics buy us? By incorporating sex and social classes in the current calculations, we get a rare view of cougar dynamics along the waterways of the Great Plains. Because of the network of interconnected rivers and streams and their dependence on and restriction to these waterways, there must have been perpetual movements of cougars up and down the riparian areas. This was not just a homogenous mix and random movement of a thousand cougars strung out along the riparian areas; they were coordinated movements of an ordered society, a society of river lions. We can envision daily movements within clusters of mostly related female cougars accomplishing their "normal activities of food gathering, mating, and caring for young." These are accompanied by the scheduled movements of territorial males, making their rounds to defend their right of access to the females. This defense is against neighboring males and the anxious, almost desperate, movements of dispersing young males, moving through these home ranges/territories looking for a chance to leave their genes to posterity. Quite a dynamic society after all.

Though cougar numbers and distribution were limited by habitat restrictions on hunting capabilities, we can still ask how many ungulates it took to fuel this society and what might have been the cougars' impact on their prey. The number of ungulates required is an easy one to calculate because the math has been done for us before. Because of the interest of the impact cougars might have on prized big game, there have been several studies attempting to estimate how many ungulates, especially deer, a cougar needs to eat yearly. These estimates have been derived from esoteric calculations of calories needed and from following cougar movements to determine when they killed something. The estimates vary widely depending on whether we're considering a solitary animal or a female with kittens, but they will give us an estimate of how many deer it took to fuel the river society.

A maximum average seems to be around thirty deer-sized ungulates per year. If we multiply the estimated number of 1,000–2,000 cougars in the plains, simple math gives us a value of 30,000 to 60,000 deer killed by cougars per year across the entire region. This may seem like a lot of deer, but this is calculated over the 2.2 million square kilometers of the prairie region. How many deer were there in this expansive area? Although we know little about the pre-settlement distribution and abundance of deer on the plains, we do know they were plentiful over most of the eastern region. Lewis and Clark in their famous expedition easily obtained the "four deer" they needed daily as they moved through the area. It appears that deer were less abundant as the expedition moved farther west, or it could be that deer were replaced by the abundant bison found there. But how many deer does "plentiful" equate to?

Today, over much of the Midwest, deer are common along the river systems because of the predator-free cover these areas provide. They also can be found out in the open rangeland, usually around the pothole marshes and swales that dot the region. Just taking into consideration the five states (North Dakota, South Dakota, Nebraska, Iowa, and Kansas) that are or were primarily grasslands, there is currently an estimated 1.6 million deer. Of this population, the annual human harvest of deer ranges from 50,000 to more than 100,000 *per state*, or around 450,000 deer killed annually in this five-state region. This means modern hunters are killing about 28 percent of the population year after year after year.

Because of the extensive amount of habitat change, we cannot directly equate the number of deer found today in the Midwest to that in pre-settlement times. Based on the above figures, however, it is probably safe to assume that over the 2.2 million square kilometers of original prairie habitat, conservatively there could have been at least two to three million deer. The taking of 30,000 to 60,000 deer per year by cougars would represent from 1 percent to 3 percent of this vast population. In comparison with the 28 percent taken by present-day hunters, the lethal impact of cougars on deer numbers was probably negligible.

The probable reason that cougars killed so few of the total number of deer is because deer likely roamed over much of the prairie landscape, feeding on grasses and forbs. In most of that landscape they were virtually free from predation by cougars; wolves and, to a lesser extent, coyotes were mostly their open-country threats. It was only when they came to the river breaks that deer faced the risk of predation by cougars. However, if those thousands of deer die each year from cougars, why would the deer come? Why not stay on the plains where they could at least see their predators coming?

As mentioned, the river systems of the plains were ribbons of lush green vegetation, forest cover, and water cutting through grasslands that at times could be dry, windy, and inhospitable. Deer came to the rivers for the water when thirsty, for the shade when hot, for the shelter when it was cold, for the fresh vegetation when winter winds dried prairie grasses. Although there may have been a resident population of deer living along the rivers, many who came were visitors to use the resources the river breaks had to offer and then return to the open plains.

Today, many of the deer in the prairie region are long-term residents of riparian areas. Freed of predation by cougars in these habitats, deer have found the riparian habitats to be safe havens from other modern-day threats. As a result, they are overusing and damaging the vegetation they depend on.[20] Because of the work by Ripple and Beschta, there are indications that this was not always the case. There was a time when grazers and browsers did not linger along streambanks long enough to do the damage they are doing today. The reason for this is fear of cougars.

This reiterates the importance of the nonlethal impacts prairie cougars had on their prey. If cougars were killing that many thousands of deer per year, they were directly scaring the 80 percent of the deer (140,000 to 240,000) that escaped their attacks. As mentioned earlier, this information gets around, and pre-settlement deer had to view river areas as desirable but highly dangerous areas to go, rivers of food and water but also of death. Places you went to, did what you needed, drank water, snatched a few fitful bites of green forage, and got out of there, before you got killed. Because of this fear, the nonlethal impact of cougars spread out along the rivers and streams across the whole prairie region, magnifying thousands of times more their ecological role than their low numbers or lethal impact would otherwise indicate.

After all this, what can we say about the ecological role of cougars in the plains region? First, their occurrence in this region was most likely specialized spatially. It was a species of the river breaks; it probably rarely ventured onto the open plains because its hunting efficiency was low and because of threats by wolves. Though it was limited to river systems, because of the extensive network of rivers and streams running across the plains, the species still probably covered a large geographical area. Over that area, its impacts were probably not felt too far away from riparian areas. Even within these areas, the lethal impact of cougars on their prey populations was almost certainly low. Indeed, deer and elk relied on the rivers for water and were vulnerable when they used these areas, but their ability to escape into the open-grass areas in all probability prevented cougars from being too effective. The biggest ecological role that the plains cougars played was perhaps in safeguarding the habitat of the riparian zones. These zones, because of their water and lush plant life, were indeed oases. They were fragile environments that without the presence of cougars would have been susceptible to overgrazing and browsing by the deer using these areas. As the cows of today idle under and destroy the trees and underbrush of riparian areas across the Midwest, so too would have these past herds of ungulates, were it not for fear of the cougar. Cougars prevented that from happening and in doing so preserved the delicate balance that maintained the vitality of the riparian habitat for all inhabitants of the prairie. Indeed, all things considered, this is not a minor role at all.

The plains of today are far from what they were when cougars roamed the riverbanks. Much of the prairie habitat has been converted into farmland. Where grasslands still prevail, herds of cows have replaced bison: the Midwest is not the wild and untamed area it once was. It is in this modified and tamed landscape that modern cougars would need to survive if they ventured out into the prairies again, as they seem to be doing. That there is sufficient food for them to survive is evident by the healthy, maybe too abundant, deer populations throughout the Midwest. However, cougars can no longer roam freely up and down the riverbanks, their travels only interrupted by towns and cities. Concerns of safety for our domestic livestock and for ourselves cloud any discussion of cougars possibly returning to the Midwest. If they were to come back, would or could they play an important role in this altered ecosystem or would they be just relics, a museum piece of what was, for people to marvel at, to pity, or, as too often, to kill on sight? It is one thing to consider the role cougars had in the original prairies, but more importantly we need to address the possibility that they could assume this role again in the modern-day landscape. Can they do it? And if so, where is it most likely to happen? Would we, while enforcing our will over the landscape, allow them to do it? Or more to the point, is there a future for cougars in the Midwest?

3

The Future of the Cougar
in the Midwest

LONG BEFORE THE LAST COUGAR WAS KILLED in the prairie states, cougars ceased to exist as an ecological force. As with the loss of other top carnivores, the few remaining individuals could not perform the ecological functions previous populations had for eons. Devoid of cougars and wolves, the prairie ecosystem stopped functioning long before much of it was turned by the plow. Cows replaced bison on the grasslands, and deer became domestic livestock in the river breaks. Where the plow did not follow the cow, continued overgrazing and "range management" altered the structure and composition of the grasslands. No longer does the endless "sea of grass" wave in the persistent winds blowing across the plains (fig. 6). The destruction of the prairie ecosystem rivals that happening in the Amazon basin in size and ecological impacts. Though many talk about restoring prairie areas, too much has been destroyed, too completely, to hope for even a minute fraction of the original ecosystem being restored. The mighty grasslands are dead.

While waves of native grasses no longer break onto tree- and shrub-lined riverbanks, corn and wheat fields still do. The Great Plains, be it covered with grass or wheat and corn, still has to drain the torrents of rain that fall on it. The Missouri, the Platte, and the Arkansas Rivers still snake their way across the heart of the plains. Fingering off these large rivers are the myriad of tributaries extending into the far reaches of the landscape. The waterways still exist; they survived because of physics and gravity. And along with them, so did the riparian habitat. Though abused, and in many places turned into towns and fields or channelized, much of the riparian areas of the original prairies persist. Plagued by flooding or just too wet and rough, these areas were never plowed.

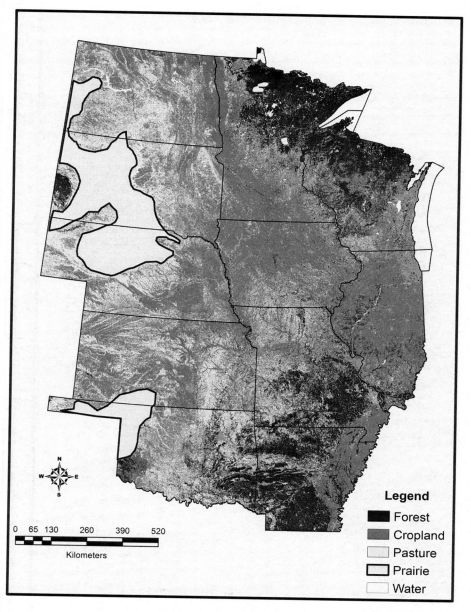

FIGURE 6. Current vegetation makeup of prairie region; note the extensive conversion of original prairie habitat to crop and pasture lands as compared to fig. 5. The outlined light-colored areas represent the vegetation that closest resembles original prairie habitat. (Based on Samson et al. 2004, modified from data available from U.S. Geological Survey, Earth Resources Observation and Science [EROS] Center, Sioux Falls, SD.)

They are battered and abused by cows and deer, but the original structures can still be found in many areas. Some are protected in county and state parks. Because we like shade and water, most parks in the Great Plains are located along streams and rivers. Much of the riparian habitat is on private land, left to grow wild when many midwestern farmers switched to growing only grains. Or, as in the drier reaches of the prairies where cows still dominate, they are left to provide shade for wandering livestock. Many streams and rivers still have extensive bands of forests protecting their shores.

Ironically, because of all these reasons, riparian habitat, the least represented in the original prairie, has now become the dominant natural vegetation in many areas. The rivers and streams no longer drain water filtered through the original grasslands, but they still collect and conduct water, and so the riparian habitat persists. And with their persistence, the hope remains for the return of their original inhabitants. In fact, it is the persistence of the riparian areas that has enabled one of the original prairie inhabitants, deer, not only to survive but to prosper. Deer populations in most plains states number in the hundreds of thousands, and most are located in the riparian areas. Interestingly, the deer seem to still follow their original movement patterns of feeding in the uplands where corn, alfalfa, and other crops have become appetizing substitutes for original prairie plants. As in pre-settlement times, deer then seek the riparian areas for water and cover, especially during fall and winter when agricultural fields are plowed under. Without the cover and native food resources of these riparian areas, deer would find it hard to survive the harsh winter winds blowing across the barren prairie soils. Given the success of the comeback of deer to the prairie waterways, we have to ask: Can the cougar, a riparian species, also return and persist or maybe even prosper?

The wolf and the bison, dependent on upland prairie grasses that are long gone, may never come back to the plains, but can the cougar? Not only does much of the original riparian habitat the species relied on still exist, but so too does one of its main prey species, the deer. Are conditions right for the cougar's return? There definitely would seem to be sufficient deer to support cougars, but is there enough habitat for them to live and hunt in? These are not just questions that *we* are trying to answer, but it seems that cougars are also attempting to seek a response.

Within the first decade of the twenty-first century, each prairie state has reported confirmed sightings of cougars. It may be that this is the century of the cougar in the Midwest. More and more cougars appear to be trying to come back to their ancestral prairie haunts. These modern-day feline pioneers, in a reversal of the earlier human mantra that spurred human settlement of the West, seem to be heeding the call "go east, young cougar . . . !" And indeed, as with human pioneers, most of them are young animals, mainly males, dispersing from established western populations. Will they make it? Can they survive in this modern version of the plains? Or are they destined to fail, their valiant attempts ending up on the evening news as they too often die in a hail of bullets when they collide with human civilization? There is no doubt that there is indeed enough prey to support them, but is there sufficient habitat for them to coexist alongside humans? Can humans, long accustomed to living without these large predators, rekindle the pioneer spirit of their ancestors? Do they have the adventurous spark to live with cougars in their midst?

In some of the plains states, because their political borders include mountainous or forested habitat imbedded within the plains, cougars can and have returned. The two noted examples are the Black Hills of South Dakota and the Badlands of North Dakota. In the actual former prairie areas, the cougars, as did their ancestors, will have to return to and survive in the streambanks. What is the potential for them there? Do states like Nebraska and Iowa still have enough riparian habitat to support viable populations of cougars? Or have these areas gotten too small, too isolated so that even with sufficient deer, viable populations of cougars are impossible?

Although dispersing cougars will eventually answer these questions themselves, here we will use science and maps to predict the biological likelihood of where cougars might be able to return to the prairie states. By return, I mean more than just by dispersing young males, but by dispersing females who could then establish breeding populations in an area. Because of the nature of cougars it can be relatively easy to have male cougars disperse hundreds of kilometers into the prairie region and beyond. However, until we get a few reluctant females to make these journeys, cougars will not establish themselves in these distant areas. When I talk about the possibility of cougars establishing populations in any given area, I refer to the number of females that an area could support. Because they are river cats, this analysis will be based on determining how much of the riparian habitat to support these females and their young still exists in each of the prairie states (see chap. 2).

What follows is a state-by-state estimate of the amount of riparian habitat that still exists and may support cougars. To minimize conflicts with humans, in this analysis I automatically exclude riparian areas in and near towns and cities. I also exclude areas too small (less than 80 square kilometers) and too isolated (farther than 20 kilometers from other areas) so that cougars would use them only for passage (see chap. 4). After that, what is left in each state is habitat that might be able to support cougars biologically. I will then follow with the sociological likelihood that humans in these areas would accept cougars in these areas. Although the prairie region is or was a continuous ecosystem that ignores artificial state lines, I have chosen to do this analysis on the basis of the states: it is those political boundaries that affected how different regions of the prairie habitat were or were not developed. It will be these same political boundaries that determine the official response of the people living there to the return of cougars. Much of this will depend on how "civilized" the area is. Is this potential habitat imbedded in farmland or rangeland? Is the land privately owned or state or federally administered? These and other considerations will help determine whether cougars that arrive to adequate biological habitat will find a sociological tolerant or hostile environment.

How much native habitat is available to cougars and what is the arrangement or juxtaposition of that habitat across each state? For this I look first at what was historically available to cougars. This analysis is based on maps that others have put together and will help to set the stage for each state relative to what they had and what they have left. To determine what was lost, I look at the current picture of the states. How much forest land, how much prairie or at least grassland, how much riparian habitat, is left in the state? For this, I rely on land use images supplied by the U.S. Department of Agriculture (see fig. 6). These images depict in 56 x 56 meter squares what type of vegetation, including crops, appears in each state. Since we have seen that cougars, even in the prairie region, depended on forested lands, how much forest and the matrix in which the forest resides becomes the most important concern.

In interpreting what I find, I use available satellite images, including Google Earth, and GIS support, but I will not perform detailed GIS-style analyses with complex algorithms. In a few states, such analyses have been conducted and where available, I refer to them. What we have found is that cougars can occur in areas that sophisticated GIS analyses might discard. And here, I attempt to look at the landscape through the eyes of the cougars. Only the cougars will eventually prove whether my cat's-eye view is more accurate than GIS models in predicting where and how many cougars might return.

Although states along the western border of the plains regions, such as Wyoming, Montana, New Mexico, and Texas, have substantial grassland habitat, they are not normally considered the Midwest; because of the mountains or forests within their borders, these states have established cougar populations. Again, the focus of this book, and so this chapter, is on the states that normally fall within the limits defined as the Midwest and that have or have had prairie habitat. Exceptions will be made when the characteristics of a given state may be important in understanding or explaining the fate of cougars in these prairie areas.

It should be an interesting and, hopefully, enlightening virtual journey across the Midwest, looking for potential modern-day haunts where the pioneer cougars striking out from the West can find shelter and perhaps even a home. In this state-by-state journey, I advise the reader to refer to figures 5 and 6 periodically to refresh their memory of how the prairie ecosystem has changed from pre-settlement times to today.

MIDWESTERN PRAIRIE HABITATS

Arkansas

Arkansas is one of the eastern border states to the prairie region and is the first exception to the rule. It originally had very little prairie and today has less than 1 percent of this vegetation type, so technically it is neither a prairie state nor a midwestern state. Arkansas belongs in this discussion because, of all the states bordering the prairie on the east, Arkansas has the greatest potential to harbor a viable population of cougars, and, as pointed out in a computer analysis, it is the one state with the most open access to the prairie region.[1] If cougars can make it to Arkansas from the west, they will find the state covered by more than 50 percent forest, and much of that is in the Ozarks and Ouachita mountains. Also in the cougar's favor is the relatively low human density (statewide: 19.8 per square kilometer, ranking thirty-fourth nationwide) and the primarily rural nature of the state.

If cougars can become established in Arkansas, from there they can build their population for further movement northeast to Missouri (fig. 7). With viable populations of cougars on the eastern edge of the prairie region, as on the western edge, these populations can provide individuals that can move out into the prairie areas. Also, and just as important, an established population of cougars in Arkansas could supply dispersers to move farther to the east where they could eventually re-colonize eastern forests.

Where could cougars exist in Arkansas? The prime region would be in the Ozark and Ouachita mountains. Of main interest to the prairie region would be the Ozarks, which run along the northern edge of the state. The Ozarks formation actually extends into southern Missouri (a midwestern and prairie state) and a little into eastern Oklahoma. Because of this, I will discuss the Ozarks as an ecological unit rather than specifically relative to Arkansas or Missouri.

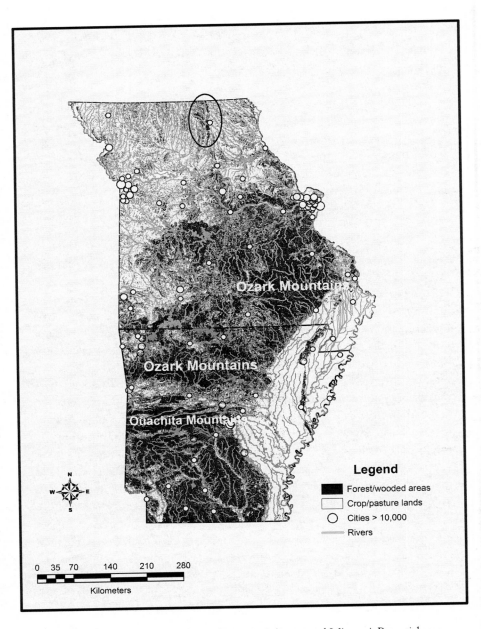

FIGURE 7. Existing forest and wooded areas in Arkansas and Missouri. Potential area where cougars could live outside of these mountains is indicated by the ellipse and discussed in the text. (Modified from data available from U.S. Geological Survey, Earth Resources Observation and Science [EROS] Center, Sioux Falls, SD.)

I include this obvious non-prairie region in the discussion because the Ozarks are the eastern analog to the Black Hills on the western edge of the prairie in South Dakota. The Ozarks could eventually function in a similar manner as the Black Hills relative to movement and persistence of cougars in prairie regions. Although the Black Hills has greater topographic relief, the Ozark complex (an estimated 122,000 square kilometers) is substantially larger than the Black Hills (approximately 8,400 square kilometers). Although much of the region in the Ozarks is private land, approximately 10,700 square kilometers is National Forest land with about 380 square kilometers designated as wilderness. Thus, this region has more than twice as much forested mountainous land under federal protection as the Black Hills (5,100 square kilometers).

The average population density of humans in the region is around seven per square kilometer. This density is less than half of the average for the state of Arkansas and similar to most of the western states (excluding California) where cougars are currently found. The U.S. Forest Service estimated that deer densities in the Ozark National Forest in 2000 were around sixteen per square kilometer. For this National Forest alone (4,600 square kilometers), this is approximately 72,000 deer. If cougars would establish at a high density of three cougars per 100 square kilometers and eat thirty deer per year per cougar, the resulting cougars (138) would annually eat approximately 6 percent of the deer population (4,150).[2] Thus, it appears that there would be ample deer to support a viable cougar population in the region.

So, all indications are that the Ozark region biologically is an area ripe for colonization by cougars, *if* they can make it across the plains (see chap. 4). Recent increases in the number of sightings of cougars in this region indicate some animals are making the journey. Would their return be sociably acceptable to the current human residents of the region (see chap. 6)? If they were to establish themselves in this region, the cougar population would function as a source for animals moving both into the plains regions of neighboring states and to the east into the vast forested region now devoid of cougars but full of deer to support them. Thus, although Arkansas is not an official plains state, its juxtaposition to the plains and its unique habitat in the Ozarks makes it a potential major player in the return of cougars to the plains.

Missouri

Because of its close association with Arkansas and containing a large portion of the Ozark Mountains, we look at Missouri next. Since it has part of the Ozark Mountains, what was said about Arkansas also holds true for Missouri.

In fact, there are increasing sightings of cougars in the southern part of the state, indicating the attraction of this area to dispersing individuals. Within the Missouri part of the Ozark complex, there is the Mark Twain National Forest (6,100 square kilometers), which exists in six parcels spread across the southern part of the state. Because Missouri also had substantial prairie areas in the northern part of the state, this potential reservoir for cougars would exist in the same political boundaries of the state. This could make it politically easier for cougars to move from the Missouri Ozarks north into the prairie region, if there is any available habitat.

Originally, approximately 27 percent of Missouri was prairie habitat, specifically tallgrass prairie. This represented a little more than 85,000 square kilometers, mostly in the northern half of the state (see chap. 2, fig. 5). Even in this prairie region of Missouri, though, only about 50 percent of the habitat was grassland. The area most dominated by grasslands was the Osage Plateau, an area of approximately 14,000 square kilometers along the central western border of the state that was nearly 70 percent grassland.[3] The rest of the region was a mixture of forest and savanna typical of the eastern edge of the Great Plains. In pre-settlement times, this mixture of open land and forest had to have been ideal hunting habitat for cougars and possibly had densities similar to those that would have been found in the Ozarks. However, because this region was also ideal for farming, cougars probably were extirpated from this area before they disappeared from the Ozarks.

Where could they live today? Today less than 1 percent of the original tallgrass prairie vegetation of Missouri remains (see fig. 6). Much of it has been converted to pasture/hay land with a smaller percent in row crops, primarily corn and soybeans. The row crops are located along the rivers and the pasture lands in the more upland areas. Much of the original forested habitat in this region has been removed, but there is an area in the north-central part along the Chariton River and its tributaries that has some larger patches of forests of 100 to 200 square kilometers, indicated by the ellipse in figure 7. These patches of deciduous trees are intermixed with crop and pasture lands, and the habitat complex is approximately 70 kilometers long north and south by 20 kilometers wide east and west—approximately 1,500 square kilometers in size. This combination of habitats is ideal for deer populations and supports some of the higher densities in the state. In addition, this type of habitat is well suited for cougars, providing cover and edge areas for them to stalk their prey (see chap. 2). So biologically this region would seem adequate to support a small population of cougars. If cougars do establish themselves in the Ozarks to the south, no doubt dispersing animals will make their way to this region. If this habitat mix were embedded in a more remote area, these animals could easily survive once they arrive. However, the open habitat is not tallgrass prairie or even open rangeland. The open areas are intensively managed pasture and cropland; the area is near to the town of Kirksville (pop. 17,000) and is likely used by local residents. Thus any cougars moving into the region would be confronted with possible interactions with humans. Whether local citizens would be willing to coexist with cougars in this region is unknown. Consequently, although biologically the area could support cougars, sociologically it is uncertain. Because of the potential reservoir to the south, whether people in this area accept cougars or not, they will have to contend with animals occasionally moving to or through the area.

As for the rest of the former plains region, there are still many forest patches associated with the various streams and rivers that run through the western part of state. They are all relatively small and surrounded by larger areas of pasture and cropland. Although there is ample deer in the region, it is doubtful that this type of habitat mix could support a cougar population. Cougars will almost certainly arrive to that part of the state but will probably come into too much contact with humans, their pets, or their livestock for residents to tolerate. Consequently, this region of the state will most likely function as a population sink where cougars will go but end up dying.

In summary, the main potential for Missouri to support cougars is actually in the non-prairie region of the state where there is ample forested habitat and topographic relief. Though this region cannot be counted prairie, it can function, as it probably historically did, as a source population for cougars in the grassland areas. From this source population, I predict individuals dispersing to the north and west of the state. Most dispersers will be young male cougars, and dispersal distances to these areas are not very long (less than 250 kilometers), so occasional dispersing females can be expected. The best chance of a small satellite population (fifteen to twenty animals) becoming established will be in the north-central region. Their persistence in this part will depend on the level of tolerance humans will have to their presence. The least likely regions where resident animals will establish are to the extreme northeast, and west-northwest parts of the state. Here the dispersing cougars will travel and either successfully pass through or end up being killed on the highways or removed because of human safety concerns. Overall, because of the Ozarks, there is the potential for Missouri to support a total population of two hundred to three hundred cougars. Most of these would be associated with the Ozarks but smaller numbers could permanently reside in some of the more forested areas to the north.

Illinois

Illinois is another eastern border state of the prairie ecosystem. Historically tallgrass prairie lands (about 34,400 square miles/13,600 square kilometers) of Illinois represented about two-thirds of the state with prairie habitat being widely distributed across the state.[4] The other third of the state consisted of primarily deciduous forest and oak savanna.[5] This forest habitat was intermixed with the prairie habitat and primarily found along the rivers and streams that drained the state. Only in the southern tip of the state did the vegetation become dominated by forest habitat.[6] Because of this mixture, Illinois had the best combination of upland prairie and river woodlands for cougars of the prairie states. This was especially true along the Mississippi and Illinois Rivers. Because of this mixture, Illinois probably had one of the higher populations of cougars of the prairie states. With Euro-American settlement, the prairie ecosystem was destroyed, and currently it is estimated that only 2,000 acres of original prairie remain. Additionally, based on USDA image data, only about 50 percent of the state's original forest lands are still present. What remains of intact forest land is primarily in the Shawnee National Forest in extreme southern Illinois (fig. 8). This forest is approximately 1,100 square kilometers divided into two parcels on the east and west sides of the state. Each parcel could possibly support around ten cougars. Such small populations by themselves may not be viable over time; however, their close association in a meta-population structure with a larger population in the Ozarks and possible future populations in Kentucky could provide sufficient interchange to maintain their viability.

The rest of the forest lands exist as scattered patches, mainly along the Illinois River and through the southern third of the state. Most of these wooded habitats are associated with the riparian areas or river breaks, as they were historically. As the name indicates, these river breaks were areas of uneven terrain, which probably saved them from the plow, so the riparian structure of the Illinois prairies seems to have been preserved to some extent. Cougars returning to these areas would find similar hunting terrain as did the previous tenants. The main difference is that these riparian forests are now imbedded in a matrix of corn and soybean fields rather than prairie grasses. As in western Missouri, biologically cougars could survive in this mixture because of the abundance of deer and hunting cover. But because of the intensively used cropland,

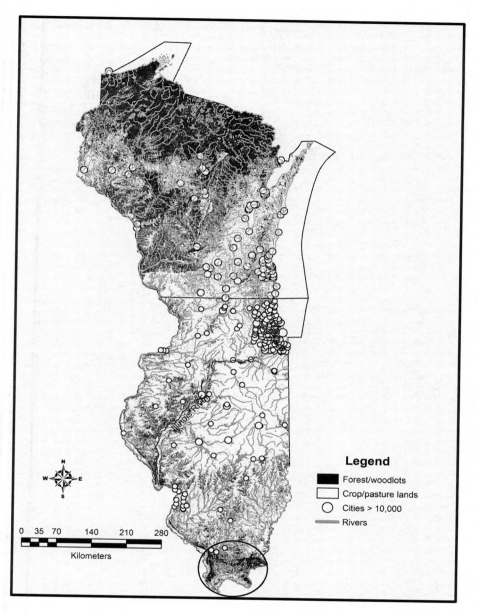

FIGURE 8. Existing forest and wooded areas in Illinois and Wisconsin. Potential area where cougars could live in southern Illinois is indicated by the ellipse and discussed in the text. (Modified from data available from U.S. Geological Survey, Earth Resources Observation and Science [EROS] Center, Sioux Falls, SD.)

cougar–human encounters would probably be too high for local residents to tolerate their occurrence there. Nevertheless, if a resident population becomes established in the southeastern Missouri Ozarks, occasional dispersing cougars may be found moving into these regions of Illinois. We anticipate that this dispersal will not be extremely high because the Mississippi River will act as a deterrent, but not a barrier, to these animals. If populations of cougars become established east of the Mississippi (see chap. 4), then movement north into these riparian forested areas might occur more frequently. Whether they survive when they get there will depend more on social factors than biological ones.

Areas north and south of the Illinois River have the least possibility of supporting cougars because they are some of the most intensively farmed regions for corn and soybeans. Here, even wooded riparian habitat is rare, and cougars would find it hard to pursue their prey. Additionally, as one moves northeast toward Chicago, the degree of urban and suburban sprawl is extensive and leaves little habitat or tolerance for cougars.

In summary, of all the eastern prairie states, Illinois probably had the most favorable mixture of prairie and forest for the existence of cougars. However, of these prairie states, Illinois lost the largest percentage of this habitat to intensive farming. Though forest habitat remains, most of it is scattered and surrounded by human activity; while cougars probably could survive in some of these isolated patches, they may not be tolerated by the human inhabitants. The best possibility of having a small resident population of cougars is in the extreme southern part of Illinois, specifically associated with the Shawnee National Forest. The level of their existence farther north would be dependent on human tolerance. Based on this assessment, Illinois would be able to support fewer than fifty cougars.

Wisconsin

Wisconsin is another prairie border state with relatively large patches of prairie lands extending only into the southern and southwestern sides of the state. Prairie habitat could be found as far northeast as Lake Winnebago and south to the shore of Lake Michigan, as well as north along the western border to River Falls. In these areas the habitat occurred in numerous patches, mixed in with oak savanna. In all, prairie and oak savanna habitat only comprised about 27 percent of the state's original vegetation.[7] Though the prairie habitat was not very extensive in Wisconsin, it probably represented a highly productive area for cougars. The mix of open prairie and oak savanna provided ideal hunting habitat for cougars in comparison to the more closed forest areas to the north.

Thus we predict that cougar numbers were probably higher in the southern part of the state in pre-settlement times.

Today in the southern portion of the original prairie-savanna habitat, most of the land is devoted to corn and soybeans. The rest is pasture land or hay crops with scattered oak woodlots. The original habitat farther north along the western border in some places is mostly a mixture of row corps and pasture. In the region north of La Crosse, much of the original prairie–oak savanna has converted to deciduous forest, probably as a result of fire suppression. In many of the areas along the eastern edge of the tallgrass prairie, fire—either natural or human caused—was an important element that maintained the prairie eco-system. In modern times, fire suppression has allowed woody vegetation to invade areas that were not converted to cultivated fields. This seems to have been the case for much of this fringe of prairie–savanna located along the western edge of Wisconsin.

Because most of the former prairie habitat in the southern part of Wiscon-sin is now intensively farmed, the potential of cougars returning to this part of the state as a viable population is low. Conversely, along the western edge of the state north of La Crosse, the forest cover has probably increased over his-toric times. Because there is still a mixture of forest and open land, there is a good chance that cougars could establish viable populations in this area (see fig. 8). Additionally, most of the northern half of the state is still covered with extensive forest habitat. In the central forest region of the state east of Eau Claire, gray wolves have returned on their own. Thus the habitat appears ade-quate to support wolves and I predict that cougars could also survive well in this vast northern region. Over the last few years, several cougars have dispersed into this area from the west and most of them still remain unaccounted for. There is also evidence that one of them, a male dubbed the Champlin-Milford cougar, appears to have made it all the way to western Connecticut where it was killed on a busy highway near Milford. More recently, others have shown up on trail cameras set up by outdoorsmen. One of them even had a radio collar on. So far all appear to be males but if a female or two make the journey success-fully (see chap. 4), it is possible that cougars could also return to Wisconsin on their own. Any population that becomes established will most likely do so in these northern regions. Once established, dispersing animals from these regions will attempt to move to the south and west into these former prairie areas.

In summary, Wisconsin will likely have a resident cougar population in the northern part of the state. Animals from this population could probably per-sist in the former prairie areas along the Mississippi River north of La Crosse and south of Eau Claire. This region of former prairie and savanna is approximately 7,000 square kilometers and so could support biologically around 70 to 120 cougars. This part of the state is not too heavily populated by humans, and so it is possible that cougars might be tolerated sociologically. Individual animals will also try to move to the south but, because of the intensive agriculture in that region, will probably not be accepted.

Iowa

Iowa is another one of the prairie states that has lost much of its native habitat. Originally Iowa was 80 percent prairie and 11 percent forested. The remaining 9 percent was mostly a mixture of wetlands and prairie-forest mix. Today, Iowa is a sea of domestic farm crops with corn and soybeans dominating 60 percent of the land mass of the state. Only 3.8 percent of the state is "grassland," much of which is not original prairie. This represents a 95 percent reduction of prairie habitat. Meanwhile, forest habitat has dropped to approximately 7 percent of the land cover or a 36 percent loss. When cougars roamed Iowa, there was almost continuous riparian forest habitat along the major rivers, like the Des Moines, the Iowa, and the Cedar, extending deep into the state. Also the northeastern corner of the state shared a continuous forested area with southwestern Wisconsin. During those times, cougars probably could easily move across this vast eastern region of Iowa. Only in the western quarter of the state did riparian forest habitat become scarce, restricted mainly along the eastern shore of the Missouri River. Because of this pattern of forested habitat, cougar numbers were probably highest toward the eastern side of the state.

Today, the Des Moines River still has extensive forest growth toward the middle of the state, but any passage to this area is blocked by the city of Des Moines (fig. 9). This large city (pop. 200,000) would be almost impossible for any cougar traveling north out of Missouri to pass without detection. The lower reaches of the Des Moines River still contain an abundance of forest patches to the west of the river. This region has a rolling topography and plenty of deer. Biologically cougars could survive here as an extension of the population that could live in north central Missouri (see fig. 7). Sociologically they would face the same problems mentioned for the Missouri population, and their persistence would depend on human tolerance to their presence. Based on the size of the area, I estimate that twenty-five to thirty cougars could live in this region.

To the northeast, there is still riparian habitat along the major rivers such as the Iowa, the Cedar, and the Wapsipinicon. In pre-settlement times, this habitat

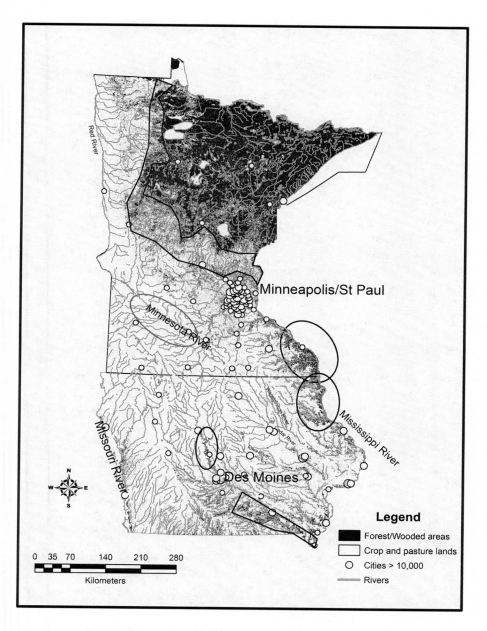

FIGURE 9. Existing forest and wooded areas in Iowa and Minnesota. Potential areas outside of the large forested area in Minnesota where cougars could live are indicated by ellipses and discussed in the text. (Modified from data available from U.S. Geological Survey, Earth Resources Observation and Science [EROS] Center, Sioux Falls, SD.)

would have been ample cover to support riverside populations of cougars. Even today, there would be enough cover and deer to support similar-sized populations. However, given that most of the surrounding landscape is intensively farmed and that there are large cities and towns such as Cedar Rapids and Waterloo, any animals venturing to these riverbanks would likely not go undetected for very long. Thus, this region would not hold much hope for supporting many cougars either biologically or sociologically.

The north-central and western parts of the state have the lowest potential to support cougars. As mentioned, originally these regions had little riparian habitat and now are essentially 100 percent devoted to croplands. Though deer can still be found in this type of agricultural landscape, the lack of cover would make it difficult for cougars to survive there biologically. This lack of cover would also make it difficult for cougars to remain hidden from humans, and so any cougar that may disperse into this region would be quickly found.

The last part of the state where it would be possible for cougars to survive is the extreme northeast corner. As mentioned, historically this was a forested region and, though reduced, still maintains numerous forest patches, especially in the bluffs and bottom areas along the Mississippi River. These remaining forest patches also connect with forest lands in Wisconsin and Minnesota (see fig. 8). Because of its distance from possible southern populations of cougars, cougars that might move into this area would more likely come from the north out of Wisconsin. If they do become established in this region, because of the denser forests, they would most likely occur along the river. There, because of the dense vegetation in this riparian area, cougars would have the best chance of living undetected by humans. Animals may move out into the forest patches farther to the east, but because of the fragmented nature of these patches, cougars would run the risk of being detected. Though a larger combined population of cougars might be found in the tri-state area, the segment living in Iowa along the river would likely not number more than ten to fifteen.

In summary, most of Iowa has some of the lowest probability of supporting cougars because of its extensive loss of native habitat. The best possibilities for small numbers of cougars would be in the south-central and northeastern regions of the state. Even in these areas, the numbers of cougars would be small and actually be part of larger interstate populations with neighboring states. The total number of cougars that could biologically live in Iowa would not be more than fifty animals. Again, in both areas, whether they could persist would be dependent on human tolerance.

Minnesota

Minnesota is Wisconsin's neighbor in many ways. Not only do they share a common border over most of both states but they also share habitats. Northeastern Minnesota and northern Wisconsin belong to the same extensive forest lands that surround western Lake Superior. As in Wisconsin, it is in these extensive forest lands where wolves continue to roam in northern Minnesota, and once reestablished, cougars could also easily make a living and, as their ancestors did, coexist with the wolves. Individuals dispersing from these forest populations to the west or south would meet varying fates as they reached the original prairie region of Minnesota.

Regarding prairies, again, Minnesota and Wisconsin both have prairie lands along the western sides of their states. They differ in how much original prairie areas they contained. While Wisconsin only had less than 15 percent of its surface covered with prairie, Minnesota was originally covered with approximately 50 percent. Original prairie habitat in the southern part of Minnesota extended to the east in a continuous manner to at least half of the state, with patches of prairie touching the border with Wisconsin. From the southern border with Iowa, continuous prairie habitat extended north to Canada along the border with the Dakotas and up to one hundred kilometers to the east into the state. Along the eastern border of the prairie in Minnesota, the grassland intertwined with tongues of northern forest kept at bay by periodic fires, forming a convoluted edge extending from the border with Wisconsin curving around the current Twin Cities to the north until Canada. This mixture of prairie and forest probably supported more cougars than the closed northern forests of the state. Although both states originally most likely had similar cougar populations in their forested regions, Minnesota probably had a greater number of plains cougars or river cats than Wisconsin. Added to this is the fact that the major rivers such as the Minnesota, the Chippewa, and the Red running through the prairie region had extensive riverside forest habitat. Coupled with the many smaller tributaries, such as the Redwood in southwestern Minnesota that also had broad bands of forest, there would have been ample riparian habitat to support populations of cougars across the region.

Today, however, most of the original prairie lands of Minnesota, like those of neighboring Iowa, have been converted to fields of corn and soybeans to the south and wheat farther to the north. Unfortunately, in some areas, the original riparian forests were also removed to make way for fields. In these areas, the hopes of cougars returning are essentially nil. Without the protective cover to hunt from and remain hidden from humans, cougars entering these areas are doomed—from starvation, visibility, and removal. The most unlikely region where cougars could survive in former prairie habitat is along the northern Red River from approximately the tri-state border of Minnesota with North and South Dakota to the border with Canada. The Red River valley on both sides of the river has been extensively converted to croplands at the expense of the original forest cover. This extensive agriculture extends eastward for about forty to fifty kilometers where pasture and grass fields begin to dominate. It is also here where some of the remnant forested areas remain (see fig. 9). Although living next to the Red River may not be possible for cougars, animals could probably survive in the band of former prairie extending to the east. Their chances of survival will increase as they near the original edge of the prairie as it turned into the predominant forest habitat farther east. This is probably the case for all the eastern border of the prairie region in Minnesota. In fact, this vast border extending from Canada and angling down toward the Twin Cities would likely provide the best combination of open land and forest cover for cougars in the state, as it likely did in pre-settlement times. As we approach the Twin Cities, human population density might become too high to allow cougars to persist, but all along this border there are rural areas where cougars likely could survive undetected. If we assume an average width of this band of 50 kilometers and a length of approximately 500 kilometers, this is an area of 25,000 square kilometers. At a density of one to two cougars per 100 square kilometers, biologically this band could support 250–500 cougars. Whether these densities could be tolerated along the total length of this band is unlikely and so the socially acceptable number of cougars would probably be closer to the lower figure. There are currently more than three thousand wolves living in northern Minnesota and so the social tolerance to cougars might be higher than in other midwestern states. Only time will tell.

The other extensive region of former prairie in the state is the southwest corner from just north of the Minnesota River to the borders of South Dakota and Iowa. I lived in this region for five years and so have firsthand knowledge of how extensively this region has been altered to cropland. In converting the farmland to large mechanized operations, windbreaks planted earlier in the twentieth century have been removed. The farming in this region is so intensive that the plow literally follows the combine. What remains for seven to eight months of the year is a barren landscape of black dirt. Winter winds wipe up the soil from the fields and deposit it as "snirt" (snow and dirt) in the ditches of the roads that crisscross the landscape at mile square intervals. This also used to be a prairie pothole region with millions of acres of marshlands and shallow prairie lakes. Many of these lakes are gone, drained to accommodate corn and soybeans. All that remains are their spirits on the topographic maps with their names and epitaph ("drained"), signaling where once lived a vibrant marshland community. As a result, much of this landscape has gone from a "country so full of game" to a wildlife desert for many species. Even the imported pheasant finds it hard to survive here; hunters prefer to cross the nearby border to South Dakota where windbreaks, marshes, and pheasants still persist.

One wildlife species that does seem to be able to prosper here is the white-tailed deer. Feeding off the corn fields in the summer and fall, deer would also find it hard to survive the harsh winters with the howling winds that race across the landscape. However, through this region run several rivers that have river valleys too narrow to plow but wide enough to support wooded vegetation. These river valleys then become the haven for deer during harsh times and in fact, human deer hunters wait along the edges of these bands of forest for the deer to come out and feed. Ironically this is probably how the river cats of the past made their living, catching deer in that narrow interface between forest and open. Since the rivers and their forest habitat remain, along with the deer, biologically there are oases within this black-soil desert where cougars could survive. The Minnesota and the Redwood Rivers, along with the Cottonwood, the Des Moines, and the Blue Earth, still provide the riparian services they gave before when the upland areas waved with tall big bluestem and other prairie grasses. These habitats are, however, not very wide or exceptionally long. And along some areas, there are major towns such as Mankato and Granite Falls. The Minnesota River is only approximately two hundred kilometers long in this part of the state and the riparian habitat probably on average is less than two kilometers wide.

Given this limited habitat, we would not expect many cougars to survive here even under the best of conditions. Add to this the likelihood that some of these animals would be detected and removed by humans, I predict that these riparian areas would support no more than ten to fifteen animals. Though this is well below any sustainable levels, cougars could probably persist in the region with an influx of individuals periodically from populations farther north. Whether even these low numbers of animals would be tolerated would depend on the density of the forest cover and use levels by people of the riparian areas. The cougars finding the most secluded places to live along these river breaks would be the ones that could persist; others, either forced into or unwittingly using areas more frequented by people, would be removed. In a sense, this is probably the way it was in pre-settlement times, with few cougars frequenting riparian areas near native villages and hunters. Those that did were likely hunted for their skins and totem spirits.

The last place to consider in Minnesota is the southeast corner (see fig. 9). Here, historically, prairie blended into savanna, then into forest along the Mississippi River. Today, there are still forested areas in this region along the river that connect with the forested areas of Wisconsin and Iowa. This band of forest is not very wide but does extend to the north to just below the Twin Cities. This area would be ideal for biologically supporting a small number (ten to twenty cougars) of river cats, which would be part of a larger tri-state population. To the west side of this forested area, farmland develops quickly, and any cougars wandering in that direction would find trouble. There is some state land (the Richard J. Dorer Memorial Hardwood State Forest) that might offer protection. If they stay along the river, they would have protection within the extensive Upper Mississippi Wildlife and Fish Refuge that runs along the river. Though small in total number, this population would remain viable because of connections with cougars to the north in Wisconsin and Minnesota, provided that cougars actually become established in these areas.

In summary, over much of the original prairie area of Minnesota, few cougars could survive, with the most likely areas being the southwestern and southeastern corners of the state. Along the long and convoluted border of the former prairie with the northern forests there may be sufficient areas where a couple hundred cougars could persist. These edge populations would be supplemented by dispersal of individuals from northern forested populations. In turn, animals from these border regions would probably maintain the viability of the small isolated populations living along the rivers farther out on the former prairie of Minnesota.

North Dakota

North Dakota is one of the first, true, completely prairie states to be considered. Originally, the state was bordered on all sides by grasslands, and except for isolated areas, most of the state was grassland. On the broadest scale, there was a band of tallgrass prairie paralleling the Red River on the east side of the state, a wide band of transition prairie to the west of that, and an even wider band of mixed prairie across the western half of the state.[8] Within these large, general categories of prairie, there were various smaller subdivisions.[9] Of importance to cougars would be the riparian habitat within the state. North Dakota has two major river systems. The first is the Red River on the eastern border with Minnesota. It is often called the Red River of the North to distinguish it from the more southern Red River along the Oklahoma–Texas border. This northern Red River is also distinguished as being one of the few rivers that flows north into Canada, emptying into Lake Winnipeg. The other major river is the Missouri, which enters toward the north along the state's western border, curves around, and leaves again about halfway across the state on the southern border. As it was in Minnesota, originally the Red River on the North Dakota side had a wide band of forest habitat. Its tributaries also had forested banks, and one of the major tributaries, the Sheyenne River, extended deep into the north-central part of the state near but not into the Devils Lake basin. The Missouri River in its original state flowed freely across the southwestern quarter of the state providing an avenue of forest habitat where cougars surely moved without restraint. Also, cougars probably had access to this vast prairie region due to the Missouri's many tributaries, especially draining the southwest corner of the state. One of the main tributaries, the Little Missouri River, drains the North Dakota Badlands region. This area, as the name indicates, is an area of rough terrain and a mixture of forest habitat, resulting in ideal conditions for cougars. Thus, though not officially mountainous or overly forested, the southwest corner of North Dakota likely had some of the higher densities of cougars in the plains ecosystem. Other notable areas that probably supported cougars were the James River in southeastern North Dakota, the Des Lacs River in the north-central region, and the Turtle Mountains just to the east and bordering Canada. All of these features provided cougars with access to a large part of North Dakota.

Overall, the river system of North Dakota ideally represented the physical structure of the plains region regarding cougars. The broad, open-grass areas were dissected by rivers of varying sizes, each with their accompanying riparian forested habitat. Most cougars in North Dakota lived along the rivers in long, linear home ranges or, when dispersing, used the river corridors, much as the native people and the early European explores did. Indeed, most of the cougars living in North Dakota were river lions. In the few areas where forested vegetation extended out and away from the rivers, such as in the Badlands or the Turtle Mountain area, cougar home ranges probably became broader and individuals moved more freely away from the rivers across the landscape.

Today, not surprisingly, much of the prairie-river habitat of the state has been altered. Unlike other states where 90 percent or more of the original grassland was converted to agricultural use, only approximately 50 percent of North Dakota is currently cropland. Approximately 38 percent of the state is still classified as some form of grassland, and about 2.5 percent of the state is covered by forest. Neither the alteration nor preservation of grasslands is uniform across the state. The most altered part of the state is the eastern third, especially along the Red River valley where farmland, including hay fields, now makes up more than 71 percent of the land cover. Grassland habitat covers a mere 16 percent of the total region. This was the original tallgrass prairie and so has the most favorable climate conditions for row crops such as wheat, corn, and soybeans. As on the Minnesota side of the river, most of the riparian habitat of the Red River has been replaced with croplands. The intensive farming in this area extends for at least eighty kilometers to the west from the northern to the southern borders and leaves little native vegetation, grass, or trees. Cougars would find it hard to make a living in this broad band of agricultural desert. Farther to the west, in the pothole region of the state, there is more grass/herbaceous cover surrounding the remaining pothole lakes and marshes. Original riverside forests have also been reduced and only small isolated patches of wooded habitat can be found. One small area is the Spirit Lake tribal lands (approximately 1,280 square kilometers) near Devils Lake. There is little forested habitat here, and these areas would not be very supportive either biologically or sociologically to cougars.

Farther west still, the cropland declines to about 38 percent of the habitat, and grassland vegetation, minus hay fields, rises to 50 percent. Here too, are the majority of the remaining forested areas originally found in the state. The most extensive area of trees can be found in the Badlands, which is protected within the Little Missouri National Grassland and the Theodore Roosevelt National Park (fig. 10). The forest cover in this area is substantially less than in those other more eastern areas discounted as being inadequate for cougars, for example Iowa and Missouri. On the other hand, the human population density in this region is much lower (less than two persons per square kilometer) than in those eastern areas (greater than fourteen persons per square kilometer), and the forested areas are imbedded in a matrix of native grasslands rather than croplands. This low density and the persistence of native grasslands results in these smaller forested areas being more capable of supporting cougars than their more eastern counterparts. Thus cougars would be able to move about the landscape with less possibility of encountering and conflicting with humans. This situation is the case for western North Dakota as well as other western midwestern states, especially where large tracts of federally administered lands exist. So for the western edge of the Midwest, less forest is needed to support cougars, and my selection criteria were adjusted accordingly.

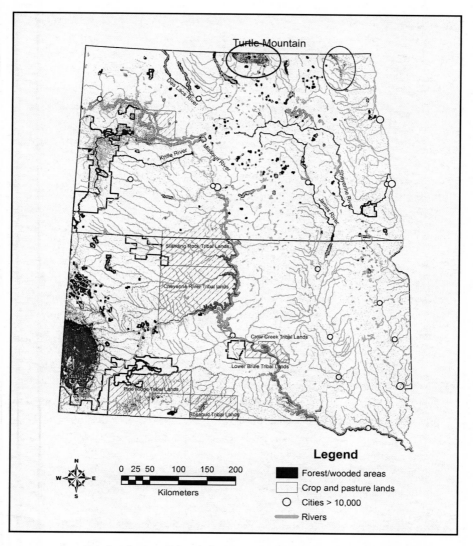

FIGURE 10. Existing forest and wooded areas in North and South Dakota (large dark area in South Dakota is the Black Hills). Potential areas outside of the Black Hills where cougars could live are indicated by ellipses and discussed in the text. Federal lands are outlined and tribal lands are indicated with diagonal lines. (Modified from data available from U.S. Geological Survey, Earth Resources Observation and Science [EROS] Center, Sioux Falls, SD.)

The Little Missouri National Grassland (4,181 square kilometers) and the National Park (285 square kilometers) were originally areas of topographic relief, grass, and trees, an ideal combination for cougars. Although there are some private land holdings within the National Grassland boundary, overall it still maintains many of its original characteristics, including the trees. This is the most likely place in the state where cougars could return and survive, and in fact they have: indeed, cougars are presently wandering the western region of North Dakota. After the last confirmed cougar was killed in 1902 (see chap. 2), there were no reported sightings until 1958. In 1991 a young female was killed near Golva, confirming that there was the potential for reproduction to occur. From that point on, sightings increased, and in 2005–6 an experimental hunting season was initiated with a quota of five cougars. The idea was to gather information on the now-evident population of cougars in the region. This hunting season has continued with the current quota of nine cougars (whether the initiation of a season on this founding population is advisable will be addressed in chap. 5). Evidently there are sufficient cougars in this region to warrant and apparently support this limited hunting of the population. With cougars in this region, it also provides an ongoing experiment as to what will be the effects of the returning cougars on wildlife populations and domestic stock herds. It will also be a test of the tolerance people will have for the return of cougars into their midst.

Just how many cougars might we expect to return to this region and where else in the state might they return to? A habitat suitability analysis done by the North Dakota Game and Fish (NDG&F) Department in 2006 identified only about 2,900 square kilometers (of the total 4,466 square kilometers in the Little Missouri National Grassland and the Theodore Roosevelt National Park) that were classified as high and moderate quality habitat. Based on this analysis and a density of one to two cougars per 100 square kilometers, we could expect to see approximately thirty to sixty animals living directly in the Badlands of North Dakota. In addition to this core area, much of the surrounding region, a mixture of farmland and grass habitat, was classified as moderate to low quality. Embedded in that region are various streams and rivers that still maintain enough riparian habitat to qualify as strips of medium to high value. One such area is along the Knife River (approximately 340 square kilometers) where the NDG&F Department identified the area as medium- to high-quality habitat. As in the pre-settlement days, it would be biologically possible for cougars, not many but some, to inhabit these smaller riparian areas. However, the upland areas do contain a high density of agricultural fields, so the social tolerance of their presence there is uncertain. Their fate there would depend on how local residents adjust to the presence of cougars in the Badlands.

One other area in this region is the Standing Rock tribal lands, which extends into South Dakota (see fig. 10). Although no cougars have yet been sighted on the tribal lands, because of the low human density and cultural views of wildlife, there is the hope that when cougars do move on to these lands, they will be tolerated by tribal members. Given all this, biologically it would be possible to have a persistent population of another twenty to thirty cougars along the rivers. Some may indeed come in conflict with human activity and be removed, but these vacancies would be filled in by dispersers from the resident population in the Badlands.

To the north of the Badlands, the Little Missouri River joins with the Missouri and flows into the current reservoir Lake Sakakawea. Around the upstream portion of the lake is another portion of the Little Missouri National Grassland and downstream from there are the Fort Berthold tribal lands (see fig. 10). The NDG&F estimated the total high-quality cougar habitat in this complex to be around 1,700 square kilometers. Cougars are also known to occur in this region with the Tribal Game and Fish Department having opened a special hunting season on them in 2007. So far, few cats have been killed, indicating that there are probably not many in the area yet. If we again assume a maximum density of one to two cougars per 100 square kilometers, this whole complex probably could support around ten to thirty cougars.

Last, there are a few small and somewhat remote locations where cougars are not known to occur yet but could survive in small subpopulations. The most likely location for this to occur is in the Turtle Mountain area in north-central North Dakota on the border with Manitoba. The Turtle Mountain complex (approximately 1,700 square kilometers) comprises forested land in the middle of a sea of agricultural fields (see fig. 10). Approximately two-thirds of this land is in North Dakota and includes the Turtle Mountain tribal lands (573 square kilometers). The area of low hills, valleys, and lakes is ideal habitat for cougars and biologically probably seventeen to thirty-five cougars could live there. The area is rather isolated from other potential habitat, and the possibility of cougars, especially females, dispersing to this area is low, but it is possible. A second, small location of forested habitat can be found east of the Turtle Mountains along the Pembina Gorge in northeast North Dakota, just west of the Red River valley. It is a small area that extends a little way into Manitoba but does have some continuous forest habitat. Based on the NDG&F report, only approximately 270 square kilometers of the area is suitable for cougars, and the report concluded that such small areas might be used only for dispersing or transient animals. I tend to agree with that assessment, although it might be possible that if a female did make it to this area, it could successfully live and raise kittens. The fate of these kittens, however, would be in doubt because of the small size of the area and the fact it is surrounded by farmland. More likely, this small area and others like it would function more as population sinks where cougars might disperse to but then be found by humans and removed.

In summary, of the states considered so far, North Dakota has the best possibility of supporting cougars on what was originally prairie habitat. This is borne out by the recent return of cougars to the extensive Badlands area of the state as well as along the Lake Sakakawea reservoir on the Missouri River. The return of cougars to this region makes biological sense in that Montana to the west has reported cougars in the eastern region of the state. The founders of the current population in North Dakota probably dispersed from this region in Montana. North Dakota, however, is not the only western prairie state where cougars have returned. As for how many cougars North Dakota could support, my estimation is probably no more than two hundred in total. Most will be around the Badlands–Lake Sakakawea reservoir and the rest dispersed in pockets of suitable habitat in the western part of the state. There are a few exceptions farther east (Turtle Mountains and the Pembina Gorge). These areas might support small, isolated populations but could function more importantly as stepping stones for cougars moving farther east (see chap. 4).

An important aspect of what is happening in North Dakota is that it is providing an example of what will be occurring over most of the former prairie states. How the citizens of the state react and deal with cougars returning will be a harbinger of what awaits cougars in other parts of the Midwest. How cougars "fit into" this region of limited habitat will then help to determine if there is room for them in other states. If cougars seek out and stay in the more remote corners of the state, avoiding interaction with people, they could be viewed as good neighbors and tolerated. But if cougars frequently enter more populated areas where they would be viewed as a threat or at least a problem, there likely will be pressure to reduce or remove them even in the areas where interactions are low. All these topics will be discussed further in chapter 5, but it is worth noting here that North Dakota is rapidly becoming one of the first testing grounds of the tolerance humans will have in their relationship with cougars in this region. The outcome of this will determine the social and political fate of cougars in the Midwest. A second testing-ground state is South Dakota, North Dakota's southern neighbor.

South Dakota

South Dakota is very similar to North Dakota. As the Missouri River leaves North Dakota, it enters and flows through the middle of South Dakota. To the east of the Missouri River, as in North Dakota, mixed prairie and tallgrass prairie dominated. To the west, South Dakota also has its own Badlands, an erosion-sculptured landscape similar to the one to the north. The main difference between South and North Dakota is the Black Hills. This landscape, rising out of the prairie floor along the western edge of South Dakota, represents the extreme eastern effort of the mountain-building forces of the Rocky Mountains. The Black Hills were so named because of their dark appearance in contrast to the light-colored prairie surrounding them. That dark color originates from the abundant forest habitat growing on this island in the prairie. Though small (8,400 square kilometers) and not terribly high in altitude (2,200 meters above sea level), the Black Hills originally contained most of the large, wild, mammal species of the Rockies. Before settlement by Euro-Americans, deer, elk, bighorn sheep, wolves, grizzly bears, black bears, and cougars roamed the forests. Most, except deer, were eradicated, with the cougars going from "numerous" in the 1880s to rare and probably extirpated by the turn of the century. Over the years, some species have been reintroduced and one—the cougar—has returned on its own. No one knows for sure when they returned, but people first began to pay attention with unconfirmed sightings of individuals. As in most cases these unconfirmed sightings were considered unreliable, but eventually they turned into confirmed ones and then actual animals being killed on the highways. An estimate in 2010 stated that around 130 adult cougars lived in the Black Hills. As such, this re-colonization of former range by a large predator is an ecological success story, unfortunately not often repeated these days. In a world where ranges of large predators are constantly shrinking, the cougar and the wolf in North America stand out as beacons of hope that those trends can be reversed. Now we need to analyze what the impact is of this success on the rest of South Dakota and the Midwest.

Cougars have returned to South Dakota, but the Black Hills, though in the middle of prairie country, is not strictly open prairie habitat. Nor is it today similar to the riparian or "badlands" open forest occasionally found in the prairies. The Black Hills is a mountainous, forested ecosystem in itself, and given its proximity to cougar populations farther west in Wyoming, it was just a matter of time before they would return. (Because this book is about cougars in the prairie ecosystem, we will not dwell on cougars in the Black Hills, apart from referring to it and its role in the plains or as an example of how people are or are not adjusting to the return of cougars.)

How about the rest of the state? What would have been the historical distribution of cougars in the plains region? Beginning east of the Missouri, this half of the state has small rivers and streams crossing the region, running in various directions and lengths. Apart from the forested riparian habitat along these waterways, there would not have been many places for cougars to hunt or be safe from wolves. Thus, historically cougar numbers would have been lowest in this half of the state. However, upon reaching the Missouri River, and points westward, this all changes. The Missouri River itself, with its extensive river breaks, would have provided free access and travel to cougars from the northern to the southern borders of the state. Additionally, major rivers such as the Grande, the Cheyenne, and the White would have given cougars access deep into the shortgrass prairie country to the west. Along these rivers and their tributaries, cougars would have traveled as far west as the Little Missouri River to the north, providing connections to populations living in the North Dakota Badlands. They could also have been able to move along the White River in the south to the Badlands of South Dakota. More significantly, though, cougars could move up the Cheyenne River to the Black Hills and in the opposite direction. Because of its connection with the Black Hills and the Missouri River, the Cheyenne River was probably the most used travel corridor for cougars dispersing from the Black Hills to the prairie areas. Moving from the Black Hills along the tributaries of the Cheyenne River and eventually along the Cheyenne itself, cougars could travel to most places in the western half of the state without leaving the banks of a river. With just a short overland trip from the Cheyenne to the White River, they had access to the Badlands with its topographic complexity and wooded areas providing excellent hunting habitat. Surely, the Cheyenne River must have been a busy place for cougars and other wildlife during those times. Add the Cheyenne to the other rivers, including the Missouri, and western South Dakota probably had one of the highest densities of cougars on the open plains of the current prairie states.

Today, as with North Dakota, the impact of agriculture differs from east to west. With the Missouri River providing a convenient dividing line, we find that east of that line, agricultural croplands dominate the landscape. The most affected areas are closest to the Minnesota and Iowa borders where more than 70 percent of the landscape is corn and soybean fields. There are some sections, mainly toward the northeast corner of the state, where grasslands predominate, but this is mostly developed pasturelands. Also, in this region there are few streams and little riparian habitat for cougars to survive. Moving west to the river, the landscape converts more to a matrix of open grassland/pastureland with embedded cropland. Although there are more open areas, again without shrub or forest cover, there is little possibility that cougars could become established in this region. In the eastern half, as in historic times, there is little habitat or space to provide opportunities for cougars to live. Later on we will see if there are opportunities for cougars to at least pass through this region to reach the forest lands of Minnesota and Wisconsin.

Before moving west of the Missouri River, we first need to consider the river itself. Historically it likely was the home of a reasonable number of river lions. Just as a rough estimate, if we consider that the section of the Missouri River here is a minimum of 550 kilometers long and the riparian habitat on average about 5 kilometers wide, it could have supported approximately fifty resident cougars. Today, most of the northern half of the river forms Lake Oahe, created by the dam near Pierre, the capital of the state. There are three other dams that form smaller reservoirs behind them. In these days the Missouri is no longer the free-flowing river it once was. How has this affected the riparian habitat? When the reservoirs formed, they inundated the closest and most productive riparian habitat. What remains are the more upland areas that would be dominated by grasslands or, where valleys lead to the river, occasional open wooded areas. Overall, the damming of rivers reduces riparian habitat and, in this case, the ability of the Missouri to support cougars. However, there still are forested areas along the river/reservoir, and in general, human use of the landscape on the western side of the river is low. The area with the most upland forest habitat is the southern section of the river just before it forms the border with Nebraska. This short section (about 70 kilometers long) would allow cougars to wander up from the river to hunt. Most of the land is open range with few cities and towns. Biologically it is possible that cougars could once again roam the Missouri. However, I would not expect more than twenty animals to be found there, mostly scattered along the river in those parts that contain forest cover.

In order to determine how feasible it would be to have cougars in the different parts of the western half of the state, we first need to consider land ownership. This is because South Dakota is the first, and probably only, prairie state that has sizable areas under federal and tribal ownership. Again, not counting the Black Hills, there are four parcels of federal land in western South Dakota. The smallest (463 square kilometers) is the Fort Pierre National Grassland next to the Missouri River. The slightly larger Grand River National Grassland (620 square kilometers) borders with North Dakota. The largest is the Buffalo Gap National Grassland (more than 2,300 square kilometers, but joined with Badlands National Park it exceeds 3,300 square kilometers). Each of these areas border tribal lands of different sizes. Together, the Fort Pierre National Grassland with the Lower Brule and Crow Creek tribal lands are the smallest (a little over 2,000 square kilometers). Though the Missouri River flows between the two tribal parcels, most of this complex is upland grassland, with little habitat for cougars. Along the river there are some forested patches that could harbor a cougar or two. How tribal members would view these additions to their land would determine whether cougars would stay.

The largest complex (around 21,000 square kilometers) is the Grand River National Grassland combined with the Cheyenne River and Standing Rock tribal lands (see fig. 10). This is a significantly large area: the Missouri River borders the two tribal lands on the east and the Cheyenne River borders on the south. In addition, the Moreau and the Grand Rivers run through the region. Although most of this area is open prairie, along with the various tributaries of these rivers, there is substantial riparian habitat to be found. Much of this region is relatively remote, and cougars using these more remote sections of the rivers could live undetected. Still, as in pre-settlement times, this riparian habitat makes up only a small portion of the region and thus this complex would likely not support more than twenty cougars. Their long-term survival in the region would depend on tribal decisions regarding the return of a native species to their lands.

Much smaller (at a combined size of 15,900 square kilometers), the complex of Buffalo Gap National Grassland, Badlands National Park, and the Pine Ridge and Rosebud tribal lands has a higher amount of usable habitat for cougars (see fig. 10). As mentioned before, the Badlands area and the surrounding National Grasslands have the topographic complexity that cougars can use. They also have scattered forest patches adequate for hunting. In addition, the Rosebud tribal lands contain forested lands along many of the tributaries of the White River, which flows through the region. Because of these characteristics, this region has the best possibility of maintaining a small population of cougars, and it is known that some already do occur in the region. My estimate, based on the topography and the amount of forested lands available, is that the area could support around twenty to thirty resident cougars. Cougars living on federal lands in this area would likely be most tolerated and have the best chance to remain. There are, however, current concerns as to the impact cougars in the Badlands National Park might have on another former resident that has been reintroduced: bighorn sheep. As we see in the current range of cougars farther west, cats are readily sacrificed to maintain dwindling sheep populations, and the concern is that this same single-species management philosophy would dominate in the National Park. As for the tribal lands, again, how well cougars would be tolerated would depend on tribal members' views concerning the return of one of the large predators that originally shared the prairies and their culture. If tribal members embrace the return of the cougar as a symbol of their sovereignty and their cultural ancestry, tribal lands potentially offer the best hope of survival of cougars in this region.

In the rest of the western part of South Dakota, much of the land consists of private ranches and small federal (Bureau of Land Management or Forest Service) holdings. Most of this land is upland prairie with the exception of a half dozen small "islands of forest" in the northwestern corner of the state that are part of the Custer National Forest. These and other small stands of forest are of limited value to cougars. Because they are embedded in fairly remote rangeland, each could support one or two female cougars with a male or two traveling among them. The only other wooded habitat would be the riparian areas of the various rivers and streams that drain this region. As in the federal and tribal lands, many of these waterways still have forested reaches, and some have topographic relief similar to the Badlands. These forested river patches, as in pre-settlement times, can be expected to biologically support a few cougars. However, over the whole region, within the private holdings, there would probably be no more than thirty to forty cougars. As for their social acceptance by the owners of these lands, many are active cattle or sheep ranchers, so their tolerance of the presence of a large predator can probably be expected to be low. Much would depend on ranchers who might suffer losses attributable to cougars. From my experience, cattle ranchers have little to worry about concerning cougar predation on adult cattle. Depending on the way calves are managed during the first few months after birth, there could indeed be concerns, which I address in more detail later.

In summary, east of the Missouri River, there is little hope (or concern) that cougars will reestablish themselves in that region of the state. Cougar numbers were probably low historically, and now with a majority of the native vegetation being converted to croplands, there is little native area where cougars could survive biologically or be tolerated politically. To the west of the Missouri, the Black Hills area now contains a resident population of around 130 animals. As is happening, dispersing animals are moving out from the Black Hills and establishing in various regions where adequate habitat exists, mostly on federal or tribal lands. On these lands, politically and culturally, cougars have the best chance for acceptance. On the extensive private holdings, cougars will move out and attempt to establish in all the forested plateaus and river banks they can find. Politically, though, they will be able to remain only in the more remote areas, as long as they don't develop conflicts with the livestock industry. Overall, on the prairie regions of western South Dakota, I predict approximately one hundred cougars living in varying sized subpopulations in a metapopulation structure. The fate of each subpopulation would depend on their size and distance from others, especially the Black Hills, and the political acceptance of their presence.

Nebraska

Nebraska is just to the south of South Dakota, and what is happening to the north of its border will directly affect the state. It is the next state downstream on the Missouri River. The Missouri forms Nebraska's eastern border, which further links it historically with its two neighbors to the north, but this is where the similarity ends. Nebraska is truly a prairie state. It does not have any extensive badlands as do the Dakotas, even though there is a small area referred to as such. Neither does it have mountains like the Black Hills, but it does have a "Pine Ridge" area, which, as the name indicates, is an extensive ridge of ponderosa pine forest. Apart from this ridge area in the northwest corner of the state, the land of Nebraska was once covered with 95 percent original prairie habitat (see chap. 2, fig. 5). As such, it had them all—tallgrass prairie in the upland areas of the east, mixed-grass in lowland and sandhills in the center, and shortgrass in the rocky gravel soils of the west. It was truly the heart of the prairie ecosystem where the grass waved unabated from the western to the eastern border. People joke that Nebraska is so open you can watch your dog run away from home for five days.

In this flat region, there were few geologic anomalies providing forested habitat for cougars to live, as we have seen in other states. Those that do exist are small, for example Wildcat Hills near Scotts Bluff (200 square kilometers) might have provided habitat sufficient for two to four cougars. The forested lands and rugged lands of the Pine Ridge region (about 1,000 square kilometers) might have provided a long and narrow home for another ten to fifteen animals. But that was it: no Black Hills, no Ozarks. If cougars were to have lived in pre-settlement Nebraska, it would have to have been mainly along the rivers. As with a typical prairie, Nebraska had several rivers, the main one being the Platte River. The Platte extended across the entire state, exiting Wyoming as the North Platte and from Colorado as the South Platte; they quickly joined together and ran free until entering the Missouri River on the eastern border. In this manner, the Platte River dissected the state into north and south, and its tributaries extended deep into the surrounding lands. Two other substantial rivers, the Republican and the Niobrara, also drained the state but more along the southern and northern borders, respectively (fig. 11).

All of these rivers contained riparian habitat. Pre-settlement vegetation maps show extensive stands of deciduous forests along these three main rivers and their tributaries. Of worth noting is that the Niobrara also had a relatively extensive stand of pine forests along its northern side and in the valleys leading down to its shore. There were even some cedar forests surrounding the Loup (North, Middle, and South) Rivers (tributaries of the Platte) in the center of the state. Finally, the Missouri River, though just forming the eastern border of the state, also had extensive deciduous forests along its Nebraska side.

It is in these forested river banks that cougars would have been found. Animals living there were bound to these rivers for their livelihood and safety, and cougars probably rarely roamed beyond the security of these wooded areas. Fortunately, rivers and streams extended over much of the state and, if forested habitat were available, so too did cougars. There were, however, large sections lacking rivers or streams where cougars probably rarely went. One of these areas is the sandhills country in north central Nebraska (see fig. 11). This extensive sandy region (50,000–60,000 square kilometers) with plentiful marshes and shallow pools contains the headwaters of various rivers and so lacked the riparian vegetation. Although the region is noted for its undulating grass-covered sand dunes, the lack of stalking cover would have precluded its use by cougars.

Because of this confinement to the rivers, the total number of cougars living in the state in pre-settlement times was probably low, an assessment reached by J. Knox Jones Jr. in 1949. Each river would have had its own subpopulation stretched out along its banks, separated, even isolated, by the prairie lands between rivers. Even within a river, the population would have been divided up by breaks in the forest patches. Each of these patches would have supported a few females while males carved out long, narrow territories big enough to make it genetically worth their time and effort. The larger patches, for example the pine forests along the Niobrara River (300 square kilometers), would have supported only three to six females and their kittens. With cougar numbers spread out and separated over these large distances, it is easy to imagine that the more distant patches in many of the smaller tributaries could become momentarily vacant with the death of the one or two residents living there. This might have

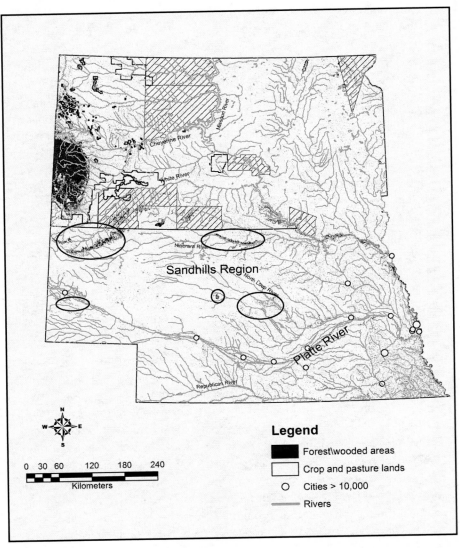

FIGURE 11. Existing forest and wooded areas in Nebraska and South Dakota (South Dakota is included because of its role in cougars in Nebraska). Potential areas where cougars could live in Nebraska are indicated by ellipses and discussed in the text. (Modified from data available from U.S. Geological Survey, Earth Resources Observation and Science [EROS] Center, Sioux Falls, SD.)

happened frequently because in these small, semi-isolated forest patches cougars were vulnerable to harassment and predation by wolves and human hunters. These vacancies would have had to be filled with little help from outside of the region. Except for an occasional disperser from the distant Black Hills, there were no large population "centers" that would provide migrating animals to rescue these smaller subpopulations. The population centers in this system would have been the major rivers, which because of the narrow limits of riparian habitat probably did not have many cougars per linear distance. Re-colonization would have had to wait until females from farther downstream rediscovered these distance patches. Such must have been the life of cougars in Nebraska, extended yet fragile. It is probably because of their low numbers and their patchy distribution that cougars were rarely mentioned and easily extirpated from the state (see chap. 1).

Today, what does Nebraska hold in store for cougars wanting to return to native lands? One might think that because of the low amount of historic habitat, returning to Nebraska would not be high on their list. Yet Nebraska seems to be one of the hotbeds of cougar sightings. Between 1991 and 2009 there have been more than eighty confirmed sightings of cougars over the state, and more continue. Where those sightings have been indicates where the cougars are probably coming from and how they are moving across the region. Meanwhile, the fate of some of these animals is a good preliminary sign of what lies in store for them as they attempt to return to Nebraska. We will put these sightings together with the current habitat conditions to consider the biological and sociological ability of different parts of the state to support cougars.

Starting with the eastern region, as we have seen in other states, the former tallgrass prairies of this region have mostly (greater than 70 percent) been converted to croplands. Also, most of the original forest habitat along the Missouri River has been reduced to a few small patches, the largest one being less than thirty square kilometers in size. Additionally, these patches are surrounded by an intensively farmed landscape. This is also the case for most of the Platte River, where there used to be almost continuous forested habitat extending westward along its banks. There still are forested stretches along the Platte, but most are narrow and the surrounding land is intensively farmed, which leaves little room for a roaming cougar (see fig. 11). The other major rivers along the southern part of the state, the Republican and the Little and Big Blue Rivers, have also suffered the same fate. So, although the eastern third of the state still has some forested river areas, they are few and far between and could neither biologically nor sociologically support cougars. This is evident in the four confirmed sightings reported between 2003 and 2009. Three of these animals, all males, were either captured or killed along or near the Missouri River. A photo of a cougar moving along the Platte River in 2009 indicates that they could still travel this river corridor undetected this far east. This holds hope that even though the Platte River may not be able to support a viable population, it still might function at least in the dispersal of cougars to the east (see chap. 4), unfortunately, however, often to their destruction.

To the northeast of the state, the farmland is less intensive as we approach the mouth of the Niobrara River. As we move upstream, this river farmland gives way to grassland, and much of the wooded habitat, including the large patch of pine forest mentioned earlier, can still be found (see fig. 11). Biologically this part of Nebraska could still support a few cougars, probably no more than ten to fifteen animals. That this area is capable of possibly supporting cougars has not gone unnoticed by dispersing animals. Along this river there have been six confirmed sightings between 2002 and 2008, and sightings in this area continue. So far only one of these animals, a male from South Dakota with a radio collar on it, was killed, by a car. Most of the sightings come from the region of the large patch of pines, but at least one is from close to the confluence of the Niobrara with the Missouri River. Because this is a slightly more remote area than southeastern Nebraska, it is possible that animals might be able to live undetected in some parts. Once detected, though, there is always the question of how the current citizens of this part of Nebraska would react. The long-term genetic survival of these isolated animals would depend on interchange with animals that could establish themselves along the Missouri River in South Dakota or that travel downstream from the more western population in the Pine Ridge area.

In the central part of the state, the sandhill country originally probably had few cougars living there. Because the sandy, unstable soil is unfit for agriculture, there are few farms there, making this one of the most remote sections of Nebraska. If a cougar could find some stalking habitat, it might be able to survive undetected. As it turns out, in 1902 stalking cover was planted right in the middle of the sandhill country (see fig. 11). In an experiment to see if trees could grow in the middle of the Great Plains, Charles E. Bessey established a large (80.9 square kilometers) planted forest. It survived and is now part of the Nebraska National Forest. From satellite images, this area stands out as a black irregular polygon surrounded by the brown-green prairie habitat. Though isolated in the center of the sandhills, at least one cougar found its way there. Its skull was discovered in 2004. This discovery is double-edged, indicating that perhaps a few cougars could live associated with this "forest" or that even with some cover, making a living on the sandhills is difficult for cougars. The other four sightings reported from this central region are exceptions that prove the rule, turning up along the rivers in the region. In 2006 a set of tracks was confirmed in the southern edge of the sandhills in the area where the original red cedar habitat grew along the headwaters of the South and Middle Loup Rivers. Three others were reported farther downstream where the branches of the three Loup Rivers join. One of these animals, a male, was killed in 2000 as it entered the small town of St. Paul along the South Loup River. Thus, as in the past, the sandhills region would still appear to be a poor place for cougars to live. Even though some of the patches of forested habitat along the edges of this region still remain, they are just too small and scattered for them to support more than a few cougars, especially if there were no human activity in the region. Unfortunately, these areas are attractive to human activity, and so both biologically and sociologically the central region in general does not hold much promise for cougars. If they do move into this region, they will exist as very small and isolated pockets of one or two females who would be highly susceptible to extirpation.

What this leaves us is the western third of Nebraska. The part of this region south of the North Platte River, including along the river itself, is devoted to cropland. The only remaining forested habitat is found in the Wildcat Hills, just south of the North Platte River. In pre-settlement times, this escarpment was small and probably supported only a few cougars. And today it seems that the scattered trees and rocky canyons are still attractive to modern-day cats. From 1996 to 2009 there have been eight sightings associated with this area. Half of them are of animals that were either trying to get to the hills or coming down off of the hills and ended up being killed in the town of Scottsbluff. One of these was a female killed in 2009, indicating the possibility that reproduction could occur in the area. The other sightings are tracks and photos from in and around the area. Because of the attractiveness of the Wildcat Hills to cougars, it is very likely that a few animals are still roaming the area. Whether these animals could be considered an established population awaits confirmation of reproduction. But, again biologically, it is unlikely that the Hills could support more than a few females, which would be sociologically confined to the rough terrain of the area. As noted earlier, wandering out of this area to nearby towns is deadly.

The last area of this region to consider is north of the North Platte River. Between the North Platte and the Niobrara River there is a mixture of farmland and grassland but little forest cover (see fig. 11). Thus, though cougars may move along the two rivers and maybe between them, it is not expected that they would establish there. However, north of the Niobrara River is the Pine Ridge area. This was a relatively large (more than 1,000 square kilometers) area of forest and canyon lands ideal for cougars. Because of its size and characteristics, it could historically have had ten to fifteen resident animals living there.

Fortunately, this area and the immediate surrounding lands to the north have not changed much. The forested areas remain, a large part of them protected within the Nebraska National Forest. Immediately to the northwest the prairie lands are protected within the Oglala National Grassland. Although cougars would not be expected to use the National Grassland, except for the riparian areas of the White River that originates there, its federal status insures a buffer of low-level human activity along the north side of the ridges and forests they would use. Thus, of all the areas in Nebraska formally used by cougars, the Pine Ridge presents the best possibility for survival of a small population of modern-day cougars. And the cougars apparently know this.

From 1991 to the present, there have been more than sixty confirmed sightings of cougars in this area, including photos, tracks, and animals killed by cars and by hunters. The sightings have been of males and females, and females with kittens, verifying that reproduction has been occurring. Although it is unknown just how many cougars might be wandering around the Pine Ridge area, this region is now being recognized as having an established population. A population of cougars in this area would be small but not necessarily isolated, as animals in the Wildcat Hills might be. A mere ninety kilometers to the north across relatively uninhabited terrain are the Black Hills of South Dakota, where a known population of more than one hundred cougars exists. The Pine Ridge formation of Nebraska also extends east into the same formation within the Pine Ridge tribal lands in South Dakota and farther north into the Badlands. All of this is good habitat for cougars, and so animals living in Nebraska are actually part of a larger population extending to the northeast. Rather than a population in themselves, they could be considered the southern portion of this larger population. Males, at least from this extended population, could surely move freely over this entire area. Females, through generational smaller steps, would also move about this whole complex, maintaining the viability and genetic diversity of the Nebraska portion of the population. Because of its connection to cougars to the north and possible link with the Black Hills, this small Nebraska population would appear to be relatively secure biologically. And because of the remoteness and public ownership of the region, it also would appear to be sociologically secure.

In summary, historically Nebraska probably did not have very many cougars wandering its predominately prairie habitat. Restricted to the rivers, presettlement cougars moved in linear fashion across the state. In modern times, the reduction of riparian habitat and the increase in human activity further reduced the rivers' ability to support cougars over most of the state. The terminal fates of cougars returning to various parts of the state attest to the probability that much of the state could no longer support viable populations of cougars. Although there are more isolated places with cover where a few cougars might survive, for example the Wildcat Hills, to say that these would be viable populations is a biological misnomer. The only parts of the state where small viable populations are possible are the Pine Ridge area in the northwest corner of the state and along the Niobrara River to the east. Because parts of the Pine Ridge area are federal lands with few people, it is here that cougars will have the fewest conflicts with humans. Though small (fewer than thirty animals), the viability of these Nebraska populations is insured because of their connection to larger numbers of cougars to the north.

This is not to say that cougars will not continue to try to move across the state. Dispersing cougars will continue to show up along the various rivers of the state as they move eastward down them. Some may even make it to the Missouri River and beyond. Most, however, will probably be detected as they have to pass along riverbanks with scant tree cover and skirt the small and not-so-small towns and cities blocking their paths. What happens to these individuals who make a wrong turn in their pioneering travels will be a test of the pioneer spirit of the people of Nebraska, and their understanding and tolerance to once again coexist with cougars.

Kansas

The next state under consideration is Kansas. Like Nebraska, Kansas is a pure prairie state. Some may consider it even more so than Nebraska: there are no mountains, ridges, or escarpments within its boundaries that could favor forest habitat. It does have an extensive area of rolling hills, the Flint Hills region that extends from the southern border to just west of Topeka in the north and was covered with prairie grasses, as was historically the majority of the state (fig. 12).[10] Nevertheless, early records seem to indicate that Kansas originally had more cougars than Nebraska had. As late as 1885, cougars appeared "abundant" in the southern part of the state and were reported as "common" in the eastern part of the state around the time of settlement.[11] Though flatter than Nebraska, it seems contradictory that Kansas might have had more cougars.

That is until we look at the original vegetation maps of the state. When we do this, we see that Kansas is probably more like Iowa in that along its eastern border there were extensive areas of prairie oak–hickory savanna as well as pure oak–hickory forest. These forests and savanna vegetation extended to about 80 kilometers or more to the west and equaled approximately 20,000 square kilometers of ideal habitat for cougars. At a modest density of one cougar per 100 square kilometers, this part of the state could have originally had two hundred or more animals roaming the countryside, probably more than what could have been found in all of Nebraska.

Farther to the west, Kansas takes on its classic prairie characteristics; any forested habitat becomes restricted to the rivers and streams of the state. Again based on historical records, at least the two major rivers, the Arkansas and the Kansas/Smoky Hill, were heavily forested along most of their courses through the state. There were indications that many of their tributaries were forested, at least in parts. So in the central and western parts of Kansas, the cougars would again be river lions, and their abundance and distribution would be limited by the number, location, and length of forested river banks in the state. Because of the normally narrow width of these riparian areas, even the longest rivers probably did not harbor many cougars. For example, the Smoky Hill/Kansas River runs approximately 700 kilometers through the state. Even if we assume continuous forest habitat in a three-kilometer band around the river, the total area is only 2,100 square kilometers. Again, at a density of one cougar per 100 square kilometers, this river would have only twenty to thirty animals spread out across the total length of the state, or three to four animals per 100 linear kilometers. More than likely there were breaks in the riverside vegetation, so the real number of cougars found along this river would probably have been fewer. As with the other states, Kansas had its own distinctive combination of vegetation and geographic features that made its unique contribution to the pre-settlement abundance and distribution of cougars in the Great Plains. In this case, it appears that cougars were plentiful in the east because of extensive savanna habitat and scarcer toward the west where only riversides provided shelter and a place to hunt.

Where oak–hickory savanna once covered the eastern quarter of the state, the grass part of the savanna has now been converted to cropland or pasture land, and the forest part has become highly fragmented. There are still areas with high percentages of forest in the northeast corner of the state, around the mouths of the Republican and Big Blue Rivers, and even along the short section of the Missouri River bordering Kansas (see fig. 12). Most areas are intermixed with farmland and close to the large urban centers of Kansas City

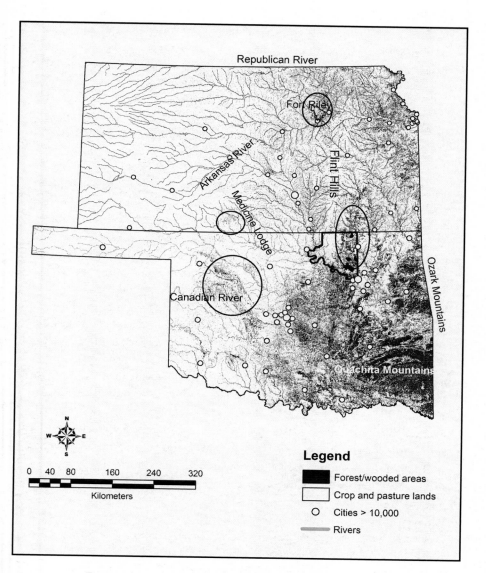

FIGURE 12. Existing forest and wooded areas in Kansas and Oklahoma. Potential areas where cougars could live outside of the Ozark and Ouachita Mountains are indicated by ellipses and discussed in the text. (Modified from data available from U.S. Geological Survey, Earth Resources Observation and Science [EROS] Center, Sioux Falls, SD.)

and Topeka. There is ample deer in this region, but because of the close proximity to these urban areas and the small sizes of the numerous fragmented
forest patches, any cougar attempting to live in these areas would easily be discovered and probably not tolerated. Thus, this would not be an area where a resident population of cougars could establish itself, although transient animals
might be attracted to the region because of the forested patches. Possibly this
is what happened when one cougar was reported in the area in 2003, making
an appearance on the wooded University of Kansas campus in Lawrence. This
report, however, was not confirmed.

One exception to this is Fort Riley, a military installation located along the
Kansas River between Junction City and Manhattan. This is an area of a little
over 400 square kilometers that has restricted access as well as some forested
habitat. Though small in size, the existence of native vegetation, including trees,
could provide a place where two to three cougars might survive undetected.
It would be extremely difficult, however, for any founding animals to arrive to
this area. Thus, the likelihood that cougars would show up at this site is very
low but, knowing cougars and their ability to disperse, not impossible.

There is one possible location in eastern Kansas where cougars might be
able to survive: along the Verdigris River on the eastern edge of the Flint Hills
region (see fig. 12). Here, as in other places in eastern Kansas, small patches
of original forest habitat persist. These patches are not large, and I would normally conclude that even biologically it would be hard for more than one or
two cougars to live there. These forested patches are connected with the extensive Flint Hills region, which because of their topography are still primarily
open grassland, mainly used for grazing. This provides a buffer on the west
side of the forest patches as well as some small wooded riparian areas within
the Flint Hills area, which individual cougars could use. In addition, just across
the border in Oklahoma are the Osage tribal lands (more than 5,800 square kilometers), comprising sparsely populated, low-impact land (8 persons per square
kilometer). Though only a small part of the tribal lands is forested, it does connect to the more forested regions in southeastern Oklahoma. As a consequence,
cougars in general will likely establish in this part of Oklahoma, and a few of
them (fewer than ten) will probably attempt to live in the forested patches of
this area in Kansas. Because of the less intensive use of the Flint Hills region,
the presence of these animals might be better tolerated than in more intensively
farmed areas to the northeast.

One other possibility of "resident" cougars in the state is in the forested
areas found in the Red Hills (Gyp Hills) region around the Medicine Bow

River in the south-central part of the state, along the border with Oklahoma (see fig. 12). As the name indicates, this area has some topographic relief and is, in fact, where the first of the two confirmed sightings of cougars has been reported, a large male killed by a hunter in 2007. It is an area of approximately 1,500 square kilometers within the state boundaries but connected to similar habitat coming up from Oklahoma, which itself connects to forested habitat along the Cimarron River. There is the potential to have a fairly large river/ forested area, of which the habitat in Kansas is the northern part, where a cougar population could biologically be supported. Because of this, there is the possibility that Kansas could have around ten to fifteen animals in this region.

The rest of Kansas to the west and north originally did not have much forest except next to the rivers. Today, this region is heavily farmed, including along the riversides, leaving little open or forested habitat remaining (see fig. 12). This region is also surrounded by intensively farmed areas in Nebraska and Colorado, and there is little possibility that cougars would even venture into this region. However, the second confirmed sighting of a cougar in Kansas came from this region in 2009, just northwest of the town of Wakeeney. Where this animal might have come from is still a mystery, but the closest animals are in the northwestern area of Nebraska. In a testimony to their stealth, it should be noted that since this cougar was sighted in October 2009, it has not been sighted again. In this intensively farmed landscape, this particular cougar has either stayed or moved on undetected to even more open country. Still, it is doubtful that cougars could permanently establish a population in this region.

In summary, although Kansas is predominately a prairie state, it still had a substantial section of dense to open forest along its eastern edge. In this area, cougars were abundant enough to be noted by early settlers. Farther to the west in the open grasslands, cougars probably were rarer and occurred only along the rivers. With the settling of Kansas, most of the prairie and forest lands were converted to agricultural use. The loss and fragmentation of the forest habitat in the east and along the riverbanks reduced the chances of modern-day cougars finding suitable places, biologically and sociologically, to live in most of the state. There are only two possible areas where cougars could reestablish small Kansas populations: the southern part of the Flint Hill region, and within the Red Hill region, both along the southern border of the state. Because of their juxtaposition next to larger remote or forested areas in Oklahoma, individuals of any southern populations establishing themselves in Oklahoma could move and live to the north in Kansas. Thus, the maximum number of resident cougars we could expect to see in Kansas would probably number less than thirty.

Oklahoma

Oklahoma is the last state to consider. Although it is not recognized as a midwestern state, I include it here for several reasons. First, much of the original habitat of Oklahoma was prairie since it is part of the Great Plains. Second, cougars were extirpated from the state in the late 1800s. And third, like Arkansas, Oklahoma has extensive forested lands along the east where a population of cougars could become established. These cougars, as mentioned earlier, could influence the movement and establishment of animals to the north. Because it is a border state, we need to look at where cougars might reestablish populations, both in the forested region and possibly farther out on the prairie regions of the state.

Oklahoma is considered a Great Plains state because the western three-fourths of the state was originally prairie habitat. Along its southern border with Texas, the southern Red River (as opposed to the one in North Dakota) is the major river running through these prairie areas, which extended into northern Texas. The other major river running through the prairie habitat in the center of the state is the Canadian River. In these prairie areas, as with the rest of the prairie habitat, cougars would be few and found along the rivers. As we move to the east, we enter the area referred to as the Cross Timbers ecoregion, which is an extensive area of prairie savanna and forest in Oklahoma and Texas. In Oklahoma, around 25,000 square kilometers of this habitat lie in a band extending from its southern border up to and slightly into Kansas. Again, this habitat is ideal for cougars, and so historically over this vast region there had to have been at least 250 or more cougars. Farther east still, Oklahoma habitat became even more forested as it connected with the Ozark regions of Arkansas and Missouri to the north and the Ouachita Mountains to the south. In this part of the state, cougar populations would have been similar to other forested and mountainous areas of the south-central and eastern United States.

With settlement of Oklahoma, many of the forested areas of the state were overharvested. Fortunately, in the rougher Ozarks and Ouachita Mountains, ample forested habitat remains. These areas are sparsely populated and, as in Arkansas, are ideal for the return of cougars to the region. Here are large national forest lands in the region, including a small part extending into Oklahoma. Thus Oklahoma could easily support viable populations of cougars in these eastern forested mountains. Again, increased sightings indicate that cougars are also attracted to this region.

The forested lands of the Ouachita Mountains connect with the original Cross Timbers region, which still maintains extensive patches of forest. Although they are scattered throughout the region, the fragments are plentiful enough to biologically support cougars. This Cross Timber habitat extends north into Kansas and is the one area where cougars could possibly live in that state (see fig. 12). Because the forest patches are highly fragmented, as with the habitat in Kansas, it is difficult to determine if cougars would be tolerated sociologically. Regardless of whether they would be, because of this region's connection to the Ouachita Mountains, once cougars become established in the mountains, dispersing animals will follow the forested patches to the north. Thus, this area will consistently receive new animals, and it is possible that in the more remote patches of forest in both Oklahoma and Kansas, small isolated populations could persist. One such area is just to the south of the Kansas border in the Osage tribal lands. This area is fairly large (5,830 square kilometers) and has a low human density (8 persons per square kilometer). Though not all the area is forested, there appears to be sufficient forest along the east edge to support a small population of twenty to thirty cougars. As discussed earlier, the sociological acceptability of cougars on these lands would depend on cultural views of tribal members.

Another area where cougars from the east could move to is in western Oklahoma. Today, much of the western one-third of the state, including the panhandle region, is intensively farmed for wheat or grazed by cattle. However, there are still scattered patches of forest to the west of Oklahoma City and along some stretches of the Canadian River and its tributaries (see fig. 12). Whether these fragments could biologically support cougars is uncertain; they are a western extension of the Cross Timbers region, and so it is possible that when cougars become established in the eastern forest habitat of Oklahoma, some animals may disperse into this more western region. With the limited forested habitat embedded in the matrix of agricultural land, however, the presence of cougars in this region may not be tolerated.

In summary, Oklahoma has an excellent opportunity to maintain viable populations of cougars in its eastern forest areas and may be on its way to having them. Though not in the plains, these forest populations could function as sources supplying cougars to areas to the west, primarily in the Cross Timbers region of the state and even farther north into Kansas. Because most of the former open prairie areas are currently farmed, it is not likely that cougars will establish themselves in these areas. But because of the source population to the east and the forest patches west of there, dispersing animals will likely be a constant characteristic of these plains regions.

PUTTING IT ALL TOGETHER

Up to now we have considered the vast midwestern region, broken into its current political units, and we have evaluated how and where cougars could live within these state boundaries. Considering that cougars don't recognize state boundaries, we need to consider the region as a whole. What is the big picture for cougars in the Midwest? What is the total number of cougars we could expect to see returning to current or former prairie habitat? How will the possible distribution of cougars look over this expansive region? How will these immigrants function on a population level? And, finally, will they reclaim the ecological role played by their ancestors?

We now have a rough estimate of the total number of cougars the prairie habitat within this region could support biologically. If we add up my estimates from each state, minus animals in forested areas such as the Black Hills, the total number is between 700 and 1,000 animals. This is less than half my original estimates for pre-settlement times, but it seems relatively high for the amount of habitat loss that the Midwest has experienced. With well over 70 percent, even up to 90 percent, of the original prairie habitat being lost, why would I not predict similar declines in overall cougar numbers? Was I being too overly pessimistic in my pre-settlement estimates or too optimistic in my current estimates? How do we reconcile this difference?

In evaluating this estimate, we need to consider two important points. First, much of the prairie that was converted to farmland was upland habitat, which cougars originally did not use. It was these abundant upland grass areas that were the easiest to plow, making forested areas of secondary importance. Though much of this forested land was cut for timber, many areas were spared or have grown back; other forested areas were on lands too rough for plowing. Some forested areas even expanded under fire suppression along the eastern edge of the prairies. Because it was these wooded habitats that the cougars require, though extirpated, much of their habitat remained. Ironically the specialized stalking behavior of cougars, which was to their disadvantage in the original prairie, could aid their return to the modern landscape. This is in contrast to the wolf, which excelled on the open plains but is at a disadvantage in the modern-day farmland landscape.

The second point to remember is that these are biological estimates of cougar numbers or the number that the landscape could support based on cover and prey abundance with minimal human influence. In fact, under these ideal conditions, my estimates would probably be on the low side simply because deer populations are most likely much higher than when the ecosystem was dominated by grass and bison. Even though today there is less forested habitat in the Midwest, it has more deer and, subsequently, could support more cougars. In the absence of humans, cougars might easily exist in much of the fragmented forest-farmland habitat along the eastern edge of the prairie. Nonetheless, humans *are* present in these landscapes, so cougars would probably avoid these areas and those who did not would be removed. In either case, these areas were excluded. So the conversion of the prairie ecosystem to agriculture in many ways reduced the number of cougars that the present-day system could support; in other ways, it did not reduce it as much as one might suspect. Obviously, the final caveat is that cougars and human tolerance will eventually tell us how close or far my estimate is from reality.

If we go with the estimate of 700 to 1,000 cougars across the current midwestern landscape, what would the arrangement be? How would these animals function as populations and would they play an ecological role in their new homes? Concerning their possible distribution, the mathematical exercise of taking the average would give us an average density of less than 1.9 to 2.7 cougars per 10,000 square kilometers. However, as we saw, cougars are not distributed by the average but are found in some areas and not at all in others. How can we best describe the general distribution of cougars across the landscape?

To best envision how cougars will distribute themselves across the Midwest, let's set aside the models and the maps for now. Any GIS model or visual analysis of a map will at best identify only the obvious areas where cougars will settle. Cougars, not having seen the maps or the models, will often defy our predictions and show up where we least expect them. Although we can try to predict, ultimately the cougars will decide where they go. And where they go will be determined by whether an area will provide what they need. What cougars know is that they need forested areas to survive. Armed with this simple search image, they strike out across the plains to seek and test all possible forested locations that might support them. In those that prove adequate, cougars will remain; if not, they will move on. In that sense cougars moving into the Midwest will be something like dust settling in a prairie home. Coming from the source populations, animals would move out across the landscape, passing through the more open areas like dust being moved by the wind. Once finding a sheltered spot, they—and dust—will settle. As dust accumulates in the large, hardly used attics and basements, cougars too, will gather in the larger, more remote forested areas and become functional populations. Some, however, will continue to move, seeking out and settling in the smaller cracks and crevices of nature. How many cougars can accumulate in these places will depend on the size of the area. How long cougars remain in these areas depends on us. It depends on how tidy, how settled, we want our environmental home to be. Just as dust is rapidly cleaned away from the more orderly homes, cougars attempting to settle in tidy, ecological homes will also be swept away. If we tidy up too much, desire too much order in our ecological home, there will be no room for dust or cougars. If we are willing to keep some wildness, some dust on the bookcases, cougars will persist in these areas and thus, in the Midwest.

What will result is a somewhat predictable pattern of cougars living in differing-sized subpopulations or a metapopulation structure all across the Midwest. Based on the previous analyses, the basements and attics where cougars are most likely to maintain viable populations will primarily be along the western and eastern edges of the region. From there, smaller subpopulations will extend, fingerlike, into the region; other smaller cracks and crevices will be scattered across the landscape like islands adrift in the agricultural sea. Because the greatest loss of prairie habitat, having been replaced by orderly farm lands, is in the former tallgrass region, it is there, too, where there are the fewest number of cracks or crevices that will harbor cougars. Each state, regardless how tame or settled, does seem to have spaces where cougars could settle, or at least try to. Some states have more areas than others, but biologically it is still possible to have cougars return to every state in the Midwest. Whether they persist or not depends on us. As we saw, some subpopulations will be completely within state boundaries while others will be shared, as will be the responsibility for conserving them.

How will these metapopulations scattered out across the Midwest function? Will they persist or die out over time? Imagined as lights in the night, we will see that the larger subpopulations will continue to shine brightly. Whether smaller cougar subpopulations continue to shine or periodically flicker out will depend on their size and how far they are from others. Some will be large enough to sustain themselves; even though they may grow dim at times, they will brighten again. For others, their survival will depend on the continual movement of individuals dispersing from distant areas. How small and how far these subpopulations are will depend on whether they continue to shine or periodically go dark, only to be relit by the arrival of dispersing individuals. There will also be individual lights that shine as animals attempt to either pass through or settle in areas where they cannot and probably should not be tolerated. These lights will not burn long, and whether they are extinguished because these animals are killed or moved to more appropriate areas will depend on social and political decisions made across this vast land. In all cases, re-colonization of the midwestern states by cougars can only occur if there is a major shift in how people accept the presence of a large predator on the landscape.

Given that cougars have and will return to many places in the Midwest, we can ask whether their return to these areas will have any ecological significance. Will they make a difference or will they just be a curiosity, a restored ecological relic? Can any of these subpopulations function adequately as a top predator in the areas they would live, especially in the smaller ones? Can we expect a return of the ecological function described in chapter 2? In the larger areas such as the Black Hills and the Badlands, I predict that cougars are currently restoring the ecological balance between large herbivores and their environment. As did the wolves in Yellowstone, more than one hundred cougars returning to these areas will again superimpose a landscape of fear on deer and other ungulates. Deer will no longer freely move about to browse and graze where they please, eating everything in their path. The presence of cougars in these areas have and will create refuges from uncontrolled herbivory that were absent for more than a century. I predict we will see a rebirth of plant diversity in these areas where rare plant species favored by deer will re-emerge on the landscape.

How about the smaller areas where only a few cougars might return? Is this a sufficient number to have an ecological impact? Or, because of the reduced and fragmented nature of the forest patches cougars would live in, are these areas beyond ecological hope? To answer that, first we have to realize that it is these remaining forest patches, no matter how small or scattered, that are the hopes of any ecological integrity such areas have. These forest patches function as oases, or arks, as the only places where many native plant species can survive in the region. Unfortunately, because these patches are small, they are probably more heavily impacted by uncontrolled foraging of deer than are larger areas. Just as we can find hidden things in small areas easier than in larger areas, favored plant species in small forest patches hold little chance of escaping the searching eye of a foraging deer. If cougars reduce the use of these patches by deer or make them more vigilant when there, it reduces the probability these plant species will be detected. In fact, their return is probably the only hope that these remaining fragments will return to some assemblage of their past diversity. Thus, even in the smaller areas, we can expect the few cougars that might return to reassume their ecological role of tending the garden of nature.

How will we know if cougars are having an effect? How do we test these predictions of restored ecological function? Unfortunately, the opportunity to test these predictions in the Black Hills or the Badlands may already be lost.

Lacking studies previous to the arrival of cougars to these areas, we cannot make direct comparisons with today. There are still possible creative ways that studies can be designed to compare with existing historical data. We hope that such studies will be done because the results should be interesting. In other areas, such as western Minnesota and Wisconsin, it is not too late. Cougars have not yet arrived, and appropriate pre-return studies could provide valuable insights into how the arrival of cougars might affect deer behavior and result in cascading ecological impacts such as seen in Yellowstone Park with wolves. Such studies are also possible in the smaller areas where I predict cougars could return. Not only is this an exciting time for animal ecologists in the Midwest, but plant ecologists should also find stimulating research. I hope that on reading this book, plant ecologists across the Midwest will take up the call and provide these valuable data.

Finally, not all the cougars entering the midwestern prairie lands will stay. We know that several have made it to the other side and beyond, so the Midwest will act not only as a place to live for cougars but as a place to pass through. Biologically there are very different requirements for an area if it is to function as a place to live versus a place in transit. If cougars are going to live in an area, the habitat will have to provide a lot more in terms of food and shelter than if they are just passing through. Obviously areas that could support cougar populations could also provide for the traveling needs of a transient animal. However, many of the areas I discarded as being incapable of supporting cougars because of their lack of habitat, small size, or close proximity to humans could easily function as way stations for overland travel of pioneering cougars. In fact, even some areas that would seem incredibly hostile to cougar survival might be seasonally beneficial to such travel. Consequently, with regards to the traveling cougar, we need to readjust our criteria and look at the landscape not through the eyes of a cougar desiring to stay but of a cougar trying to pass through to points farther along. What does the traveling cougar need? What are the hazards, the road blocks in its way? Can it make and if so, where?

4 To the Prairies and Beyond

ALTHOUGH WE HAVE IDENTIFIED A WIDE RANGE of areas where cougars could actually survive within the plains region, not all the cougar pioneers striking out from their homelands in the West seek or find a place to settle in the plains area. As with many of the western-moving human pioneers of the past, reaching the plains was not the goal but just part of the journey. Many human pioneers looked at the plains region as an area to cross as they headed toward what they hoped were richer lands in the West. So, too, some of the cougar pioneers striking out across the plains have no intention of staying but, driven by other forces, continue to head east, in hopes of finding better habitat. For many, the main force driving them is the reproductive urge. Most of these pioneers are young males who, under normal conditions, usually disperse from their home areas looking for possible mates. Within the current range of cougars, available females exist, and many young males only go as far as they need to establish their breeding territory. For males moving east, females can be few and far between. Once a male strikes out across the plains looking for mates, he is unwittingly committing himself to a longer journey eastward than he could possibly foresee. Every day, he moves in anticipation of fulfilling that goal of finding females. Whether he will succeed, regardless of how far east he travels, is another issue. Can he continue to move east, and if so, how will he move across the plains?

One of the most recent exciting yet tragic developments concerning cougars in the plains region is the growing number of cougars who are striking off across this great inland sea of grass, wheat, and corn (see chap. 3). As with the early human explorers from the East, these adventurous souls are heading off to lands unknown, without aid of map or compass. This is exciting because these hardy individuals are the vanguard of a hopeful re-colonization of ancestral grounds not just in the plains but also to vast areas farther to the east (see chap. 6). It is tragic because, as with most exploratory thrusts into unknown lands, many perish. They find themselves in an alien human world that has little tolerance for their presence. Confused and scared, they end up in trees, under porches, in backyards. People, untrained and inexperienced with dealing with cougars, react with fear, and the first response usually results in the disastrous end to the cougar's adventure. Most do not make it very far. Of twenty-nine dispersing cougars between 1990 and 2006, twenty-one dispersed less than four hundred kilometers to the east before being spotted and killed. Seven of these never made it more than one hundred kilometers. One hardy individual made it all the way from the Black Hills of South Dakota, across one of the most heavily farmed regions of the upper Midwest, to the safety of the forests of Wisconsin, only to die in a hail of gunfire in a Chicago suburb. Even if they make it across the plains, it is no assurance that they will survive. Thus, the Great Plains is both a conduit and a barrier for eastern movements of cougars. Just how permeable that barrier is will determine when and if enough colonizing cougars can make it to the eastern forests. This becomes important because, barring radical changes in political will, re-colonization of more eastern states will have to come from these wandering migrants venturing out into the open plains. What happens to them once they reach these forests is anyone's guess. Our concern here is to try and evaluate the *probability* of cougars actually making the journey toward the East, and where in the current plains landscape they have the best chance of this happening.

Consequently, besides assessing the capability of the prairie states in supporting viable populations of cougars, we need to evaluate the capability of these states in facilitating the movements of cougars across their lands. Even though a prairie state may not be able to support populations of cougars, it may assist or impede their passage to more adequate habitat. We need to move beyond state boundaries and evaluate the total landscape for potential corridors. Habitat and cougars do not recognize state limits, and we need to look at the landscape through the eyes of a cougar heading off into the east from known populations along the most eastern edge of their current range.

LaRue and Nielsen in 2008 did just that in their in-depth GIS study of possible movement corridors across the prairie. Their analysis included factors such as land cover, human density, slope, distance to paved roads, and distance to water. With this analysis and information on possible trajectories of known dispersing cougars, they determined what the least-cost path (LCP) would be across the Great Plains. Theirs was a superb first view of how to get across the plains, or not, with minimal contact with humans. It was, out of necessity, a broad-scale approach, but still they were able to demonstrate several cold, hard facts about trans-prairie travel. First, even though over most of the western half of the prairie region there are few people and few developed lands, because of the open nature of the area, it makes being spotted much easier than in forested areas. This explains why so many dispersing animals only make it a short distance before being spotted and killed. Second, from almost the northern border of the United States down to the Gulf of Mexico, there is a relatively wide band of densely used farmland. Thousands of square kilometers of intensively farmed land leave little cover for a wandering cougar. Not surprisingly many of the rest of the dispersing cougars' paths abruptly end upon reaching this wall of hostile terrain. As dim as the picture may seem, LaRue and Nielsen did find a ray of hope. Based on their LCP analyses, they predicted that the best chance of a cougar making it to the eastern edge of the prairie was actually through northeast Texas. Even though the Dallas–Fort Worth megalopolis dominates a large portion of the landscape in that area, there is little development to the west, and a narrow band of semi-broken forest exists between that area and the sprawling Oklahoma City complex to the north. On the other side of this gauntlet are the forests and mountains of the Arkansas Ozarks. Indeed, as described in chapter 3 and observed by additional work of LaRue and Nielsen, western Arkansas provides a forested mountainous terrain similar to much of the habitat currently occupied by cougars farther west. The validity of their analysis is substantiated by several cougars actually having been found, and unfortunately killed, in northern Arkansas and southern Missouri. Thus, successfully crossing the Great Plains may not ensure survival on the other side. What we can hope is that other, yet to be discovered, cougars have made this journey and will be the founders of the first established cougar population east of the Great Plains in more than 150 years. Only time will tell, but the increased sightings of cougars in Arkansas and Missouri provide us with some hope.

It would seem that the best hope for eastern dispersal of cougars across the Great Plains is the southern route. Yet at least more and more cougars from the north made it through the farm belt of the Red River valley to the forests of Wisconsin. One ended up dead in Chicago; another, treed in western Wisconsin in early March 2009, escaped attempts to tranquilize it and is, hopefully, still wandering around in the Wisconsin forests. Another, in the winter of 2009–10, also made it across the northern plains to east-central Minnesota and western

Wisconsin. This cat also headed to the North Woods of Wisconsin. This particular cougar, however, did not stop there but it seems continued its incredible journey eastward, mostly undetected, as far east as western Connecticut. Since then, the increasing numbers of cougars being sighted in Minnesota and Wisconsin indicate they can make it, and again, hope rises that more, including females, will make the journey too, and will survive.

Though the best route seems to be to the south, cougars coming out of the Badlands of North Dakota and the Black Hills of South Dakota don't know this. Without the benefit of satellite images and GIS technology, they will continue heading due east. Are most of them destined to be doomed, smashing up against the seemingly impermeable wall of corn and soybeans? Where did those few that did make it pass undetected? If we can look for the holes in this wall, maybe we can enhance these pioneers' chances of surviving the trip. And are there other holes in the wall farther south for those cougars coming out of Wyoming and Colorado? After their long trip across western Nebraska and Kansas, can they slip through to forested lands in eastern Missouri and maybe points farther on?

We need to take a closer look than what LaRue and Nielsen did. Instead of general landscape features (percentage of habitat or density of roads), we need to look at specific pathways cougars might be able to take. Again, without the benefit of satellite images, cougars can't plan their trip based on what might be a favorable percentage of forested habitat fifty kilometers in the distance; they, beforehand, do not know the least-cost pathway. They make their trip one step at a time, choosing their path based on what lies ahead no more than a few hundred meters. It is this close view perspective, "through the eyes of a cougar," that we must attempt to do, to see where they may take the right or wrong turn in their perilous journey. However, unlike cougars, we can do it with the aid of our satellite eye in the sky. It is this type of virtual journey, step by step but with the aid of Google Earth, I would like to take us on—not just in one area but across the length of the Great Plains in hopes of identifying actual pathways that run the gauntlet from the eastern edge of the Rockies to the eastern edge of the prairie. It is a dauntless task to be sure, but one that we need to attempt if we are to try to help cougars in their journey.

How do we do this? As did LaRue and Nielsen in 2008, I will rely on the extensive satellite information compiled by the U.S. Geological Service (USGS) seamless map server site. Through the diligent efforts of the USGS as well as NASA and other agencies, we have incredible images of our country that show vegetation and human inhabitations. If any of you have ever used Google Earth you have benefited from their efforts. In this particular case, I downloaded images that had a resolution of thirty meters (approximately 30 yards). This means that vegetation or other structures thirty meters or so long will show up on the map. From such maps, we are able to see the patterns of native and human-altered vegetation, shapes and sizes of towns and cities, topographic features, and much more in fine enough detail to identify where cougars should go and what areas they should avoid. Additional maps or, in GIS (geographic information systems) lingo, layers, show us the locations of rivers, roads, and other landscape features. They even have a layer that shows what the United States looks like at night (fig. 13). Since cougars travel at night, this visual helps us avoid brightly lit and thus highly populated areas. If we overlay all this information, we can assess, as did LaRue and Nielsen, what the risks are for a cougar across this landscape. Going beyond LaRue and Nielsen, here I attempt a finer scale analysis to identify specific pathways cougars should take, if they had Google Earth. It will be a virtual foot journey across the plain, a trip where we will make turns or go straight based on what we would see directly ahead of us if we were walking. To do this, we selectively remove areas hostile to cougars: cultivated lands and towns/cities. What remains is a map that shows in black and gray the pattern of forest (few but often critically located) and native vegetation. From this image we can then start our journey.

The first thing to ask is: What would a cougar look for when it starts its journey? Knowing what we do about cougar habits, we do know they are a species of cover (see chap. 2). They use it not only for hunting but also instinctively for protection from the weather and from other predators, including humans. Historically, in the plains the thickest cover was found along stream and river banks and we know that ancestral plains cougars were creatures of the rivers. There is no reason to think that these new pioneers striking off across the open plains would be any different. So cougars moving out onto the plains would instinctively seek out these riparian areas, the areas of their ancestors.

That is how we will start the journey across the Great Plains, seeking the cover of a riparian area and staying with it as long as possible, as long as it heads east. Although the Great Plains region is covered with streams and rivers, many are feeder streams that eventually join into the five major rivers that cross the plains (the Missouri, the Platte, the Arkansas, the Canadian, and the Red) and then join with the Mississippi. Any cougar starting its journey from one of the smaller rivers at the western edge of the prairie would quite quickly join its fate to the flow of one of these major rivers. The question then becomes: How adequately do these rivers provide cover for movement through the plains

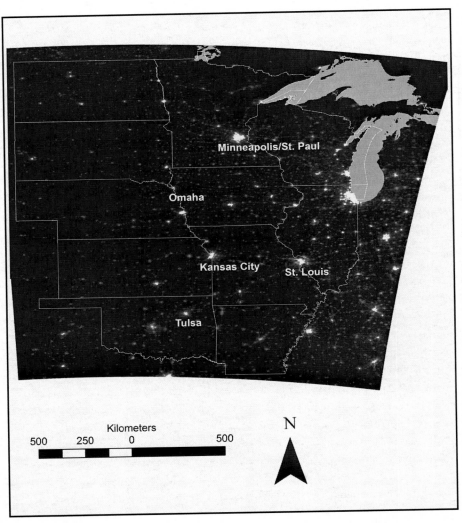

FIGURE 13. Night lights on the plains. Major cities on rivers such as Omaha shine brightly and indicate barriers to cougar movement even under the cloak of darkness. (Spatial data from U.S. Geological Survey, Earth Resources Observation and Science [EROS] Center, Sioux Falls, SD.)

region? We know that most do not contain the same riparian habitat that existed before colonization of the plains (chap. 3). Many fertile areas have been cleared for farmland, and most major prairie cities are built along these rivers. Like so many other areas, riparian habitat has become fragmented, leaving gaps between stretches of native vegetation. Where are these gaps? How big are they? Are there reasonable ways around them? These are some of a multitude of questions a cougar faces in its journey and has to answer, without prior knowledge of what lies ahead. This is why it is important for us to take the bird's-eye view and see if we can identify where a cougar, moving instinctively but unaware of what lies ahead, might make a wrong turn, might get into trouble that would abruptly end its trip. If we plot a course that a cougar would most naturally follow, we can identify the dangers ahead and, if possible, reduce or remove them, or provide a natural detour a cougar might follow. From this exercise, we can not only map the safest course through the Great Plains but also visualize one a cougar might instinctively, maybe with a little gentle prodding, follow to safety.

The Northern Routes

Armed with our GIS image tucked under our arms, let's begin our trip. First let's start with the northernmost river, the Missouri. We know that cougars can make it as far as the Badlands of North Dakota; a breeding population exists there (chap. 3). In fact, it is from this population that the half dozen or so dispersers moving across North Dakota came.[1] So we will not worry about the Missouri upstream from the North Dakota border. When we look at the habitat map, we can appreciate why there is a breeding population in the Badlands. Besides the area containing two relatively small patches of the Theodore Roosevelt National Park, these patches are imbedded in the Little Missouri National Grassland. This is a relatively large tract of native vegetation (4,200 square kilometers) that also contains substantial amounts of forested habitat and the Little Missouri River running through it (fig. 14). There are also no major roads through the area except Interstate 94 toward the southern end and only a few small towns along the border, so the area is well suited for cougars. This then is a nucleus of any easterly dispersing cougars. As it turns out, leaving the Badlands to the east is relatively easy and safe. The national grassland extends almost to the Missouri; and all along the Little Missouri to where it enters the main river it is surrounded by grassland and forested habitat. Starting the journey seems easy, and as a cougar I would feel pretty lucky to have so much native vegetation, including trees, to shelter my trip. However, the

Missouri River is wide at this point, called Lake Sakakawea, a reservoir formed by a dam about fifty kilometers downstream. Cougars can cross water but they are not big fans of doing so. As a result, cougars coming out of the National Grassland on the south side of the Little Missouri would likely stay on that side while moving east. Those that might come along the north side of the Little Missouri would become hemmed in where it joins the big river/lake. From there they would head upstream and probably cross the Missouri a few kilometers west where it narrows. Evidence for that are the several dispersing animals who were killed to the north of the Missouri, one making it as far as the U.S./Canadian border.[2] Looking at the map, we can see why this happens. These northern travelers probably head up rivers like the White Earth, which begins with ample grassland and some forest cover. This soon gives way to a predominance of farmland and an increased risk of exposure. For most of these dispersers, starting out on the wrong side of the Little Missouri is probably a sure death sentence.

For those taking the south side of the river, unfortunately major trouble also lies ahead. You don't have to go very far down the Missouri before native grassland vegetation gives way to farmland, and the forested habitat becomes patchy strips along the river's shores. Fortunately, there are still few towns in this area, and it is only about thirty kilometers before we reach the place where the river abruptly turns south. I have found that cougars can travel in excess of thirty kilometers in a day, mostly at night, so a cougar could easily cross the area of farmland during the night to a large patch of native habitat with trees just south of the dam that forms Lake Sakakawea. It is a small area of around forty square kilometers but consists of a series of forested draws running down to the river, ideal habitat for a tired cougar to spend a day or two undetected. The question then becomes: Where do I go from there? If my goal is to continue heading east, I have to cross the river. If I want to follow the river to the south, I have to travel more than seventy kilometers through open fields and pass by three small towns before I reach—perhaps—a reasonably large patch of native vegetation and some riparian woodlands. Again, this trip is possible but difficult. Unfortunately, if they do make the trip, just thirty or so kilometers to the south is the large city of Bismarck. Although the actual center of Bismarck is on the other side of the river, the urban influence extends across the river to Mandane. Running this gauntlet, even at night under the intense urban lighting, makes it another literal dead end for any cougar traveling that way. The cougar killed in north Bismarck in November 2009 is testimony to this. However, there is a slim possibility a cougar could skirt the city lights to the

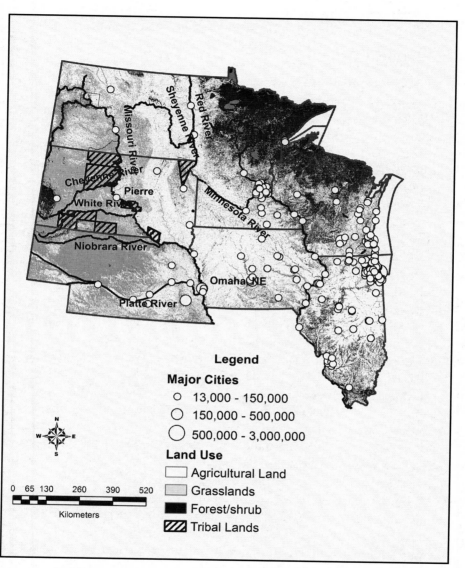

FIGURE 14. Patterns of forest, grassland, and agricultural land along major rivers running through northern plains states. Cities with more than twenty thousand inhabitants are indicated, and those mentioned in the text are labeled. (Modified from data available from U.S. Geological Survey, Earth Resources Observation and Science [EROS] Center, Sioux Falls, SD.)

west where some native vegetation still exists. If it does make it past Bismarck, conditions improve somewhat with some forest cover and expanses of grassland vegetation to the border of South Dakota. The fact that no cougars have been documented (i.e., killed) using this route indicates that Bismarck may present too formidable of a barrier, meaning the only real hope for a dispersing animal is to cross the river and head east.

To the east just north of Bismarck, there is a region of extensive grassland habitat extending more than 150 kilometers. The region is interspersed with some farmland, but this is rangeland turning into mainly marshland and small prairie lakes. There are few towns, so a cougar could move eastward during the night and rest in tall marsh vegetation during the day. The evidence of at least two cougars moving in this direction supports this contention.[3] However, not too far east the cougar runs into the agricultural wall. Starting around the James River valley and extending to the northern Red River of the North (between North Dakota and Minnesota; the river empties into Lake Winnipeg and eventually finds the Arctic Ocean), we begin to see more and more farm-land (see fig. 14). In the first 100 kilometers, until we meet the Sheyenne River, we can still find marshland habitat that may offer refuge to a passing cougar. But from there to the Minnesota border and beyond is more than 150 kilome-ters of some of the most intensively farmed land in the nation. Even though this area is sparsely populated, there is just no cover for a cougar to pass through in this region for at least ten months of the year. The only hope for a cougar to pass through this region to the forest lands of Minnesota is to move through during July and August when corn is "as high as an elephant's eye" and over 50 percent of the landscape is covered with corn, wheat, and soybeans (fig. 15). All of these crops, as well as others such as sunflowers, can provide shelter for a traveling cougar. Many of the fields in this region cover one complete square mile and are separated from adjacent ones only by narrow dirt roads. While inside a corn field, or even a wheat field, a cougar could easily travel the length of it undetected and maybe even catch a deer that frequents these fields. A brief jump or two will immerse it again into the shelter of the next mile-square field, and so on across the region. Ironically, for a brief period in mid- to late summer, the agricultural wall becomes quite porous and may actually aid in a cougar's movement to the east. Timing, however, becomes important. As with the human pioneers planning their departure from St. Louis to avoid winter snows in the mountains, cougars only have a certain window of oppor-tunity when leaving the Badlands to make it when crops are still in the fields. Unfortunately, unlike the western immigrants, eastern feline migrants do not

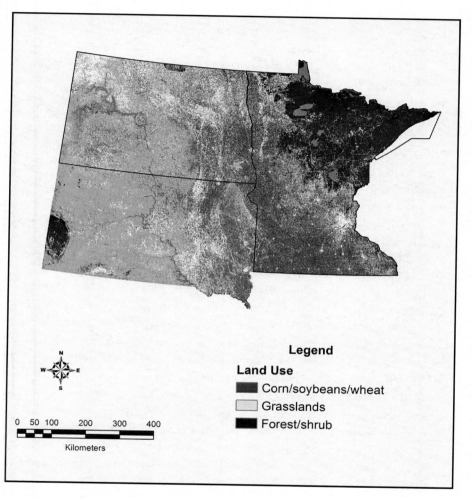

FIGURE 15. Corn, soybeans, and wheat provide cover for the movement of cougars across what would otherwise be a barren landscape as depicted in fig. 6. (Modified from data available from U.S. Geological Survey, Earth Resources Observation and Science [EROS] Center, Sioux Falls, SD.)

have the knowledge of prior travelers to inform them of when to leave, and it is a matter of chance if they arrive at an opportune time or not. The young male cougar killed near Bemidji, Minnesota, in September 2009 might have been an individual that crossed the wall in July or August while the landscape was covered with corn. It is too bad that cougars could not plan their dispersal to coincide with peak growing season in the prairie farmland.

From the badlands of North Dakota, cougars can and have successfully traveled across the landscape as far as the forests of Minnesota and beyond. Based on my analysis, this trip includes following the Missouri River until just north of Bismarck where animals then strike out across the prairie pothole region. From this point, how far a cougar gets depends on the time of year. When croplands are plowed and barren, from October to early June, there is little hope a cougar can move across the more than two hundred kilometers of open farmland. The agricultural wall, however, becomes more permeable in mid- to late summer, and cougars might be able to safely pass through the kilometers of cornfields. The evidence indicates that cougars from the North Dakota Badlands take this route, but in at least one instance a male cougar from the Black Hills of South Dakota (killed in northwest Minnesota in 2004) probably followed this route as well. In April 2008, in a hail of bullets, another Black Hills cat was killed in Chicago. This cat was seen in January in southeastern Wisconsin. Given travel times, it is possible this cat also passed through the Red River valley during late summer. Further sightings of two cougars in Wisconsin in September 2011 also indicate they are passing through the protective cover of unharvested cornfields. This then turns our attention to the next point of departure for eastern moving cougars, the Black Hills.

In the Black Hills, the picture improves tremendously. First, the recently reestablished population in the Black Hills is larger (approximately 130 adult animals). This provides many more possible dispersing individuals per year. Also, there are several favorable directions that dispersing cougars can choose from. The first is to the north. The Little Missouri River and its accompanying riparian habitat extend upstream into extreme southeastern Montana and northeast Wyoming to within twenty kilometers of the Black Hills complex. All along this stretch from the National Grassland to its approach to the Black Hills, the Little Missouri is surrounded by a broad band of grassland vegetation. There are few towns, and cougars could pass, and have done so, to the north undetected. It is probable that the founding population of cougars in the Badlands of North Dakota followed this route. Once in the Badlands, these cougars can continue their easterly journey.

Another possible exit point from the Black Hills would be directly to the east. There are several rivers and creeks, the Belle Fourche, Elk Creek, Box Elder Creek, Rapid Creek, and Spring Creek, that feed directly into the Cheyenne River, which connects with the Missouri River. Unfortunately, the towns along this route, Spearfish, Deadwood, and Rapid City, can make these exits dangerous, as evidenced by the various cougars that periodically show up in these areas. However, there are enough of these riparian areas that are surrounded by grassland habitat that cougars could easily travel to the Cheyenne River. Once to the Cheyenne, they can safely continue their journey to the Missouri (see fig. 14). The lack of cougar reports along the Cheyenne but farther to the east indicates the safety of this leg of the journey.

Once reaching the Missouri River, cougars have three choices: follow the river to the north, follow the river to the south, or strike off directly to the east. Direct eastward movement is blocked because at the conjunction of the two rivers, the Missouri again has been impeded to form Lake Oahe. Because of this impediment, the direction a cougar would head may depend on which side of the Cheyenne River it is when it reaches the Missouri. If a cougar approaches from the north side of the river, it is committed to following the Missouri northward. Fortunately, much of the north side of the Cheyenne river lies in the Cheyenne River Tribal lands, as does much of the west bank of the Missouri, including the Standing Rock Tribal lands to the north. Within these tribal lands there appears to be ample grassland vegetation and occasional forest cover all the way to Bismarck. So, unless the cougar crosses the river (about 120 kilometers north of the Cheyenne confluence) and heads east, it will suffer the fate of the 2009 animal on the streets of the city. Fortunately, just south of Bismarck, there is ample grassland vegetation leading to the same marshland/lakes complex animals from the north could move in. From that point, these southern animals likely would share similar pathways, and fates, as animals coming out of the Badlands of North Dakota.

If an animal approaches the Missouri from the south shore of the Cheyenne, again, there is ample grassland vegetation along the Missouri. A cougar could cross the river just below the dam for Lake Oahe except for the city of Pierre, the capital of South Dakota. Though this may present a logistical problem for cougars traveling this way, there is abundant grassland habitat to the west of Pierre, and cougars could skirt the city easily and continue their travels south and possibly east. Movement to the east, however, is inadvisable; not too far south of Pierre the wall begins again and in this case extends for more than 550 kilometers through southern Minnesota and northern Iowa until reaching forest lands in western Wisconsin. There are few river systems running west–east to follow, and very little native cover. The only major river system they could possibly connect to would be the Minnesota River, more than two hundred kilometers away. The chances that a cougar could make this journey undetected, even under the cover of mature corn fields, is essentially nil. Fortunately, none seems to have tried it so far. I would guess that if any do turn east, the open landscape visible to the distant horizon should discourage them. So it is back to the Missouri and to the south. Although most of the surrounding landscape on the east side of the Missouri is dominated by cropland, the west side still has extensive grassland habitat and even some forest cover for a wandering cougar. There are also two areas of tribal lands, Crow Creek and the Lower Brule, that offer habitat to rest and to possibly stay (chap. 2). This cover holds until the Missouri begins to be the border with Nebraska where the surrounding countryside on both sides of the river turns into predominately cropland. By the time the river reaches the three-corners area of South Dakota, Nebraska, and Iowa, little native vegetation is evident near the river itself. Because native vegetation along the river all but disappears at this point, few cougars would continue farther. If they did, they would run into Sioux City, Iowa, which, with South Sioux City, Nebraska, surrounds the river with urban habitat. This effectively blocks any further travel along the river itself. If a cougar chose to skirt the area along the high banks on both sides of the river, there is some forest cover available, especially under the cloak of darkness. Any success passing Sioux City, however, is short-lived as a south-moving cougar approaches Omaha. This metropolitan area is so large that it totally engulfs both sides of the river (Council Bluffs on the Iowa side) during the day and at night. This makes passing through this area almost impossible, as attested by the male cougar in 2003 who tried it and is now living in the Omaha Zoo. Patches of grassland and forest skirt the area, but they are few and small, increasing a cougar's chances of being spotted. Its only hope would be to travel approximately thirty kilometers overland southwest through mostly open farmland to the North Platte River. The discovery in 2005 of a male cougar killed on Interstate 80 near Gretna, Nebraska, just west of Omaha, indicates such a choice can be deadly. If they do make it, then they might meet up with cougars moving down the third dispersal route to the east, the North Platte River. Let's look at their movements before we consider them jointly, with the cougars coming in from the north, and continue the eastward journey together.

    The headwaters of the North Platte and Platte Rivers extend well into Wyoming and Colorado and into current cougar country. These rivers can probably act as corridors for cougars heading east and reaching western Nebraska. Most cougars moving farther down the North Platte today likely originate again from the Black Hills in South Dakota. The south side of the Black Hills slopes into a broad expanse of grassland vegetation. There are few towns and roads, and a cougar could easily move through this area unnoticed. Some individuals might connect with the White River in South Dakota and head east with no problems through the Pine Ridge Tribal lands. Others, heading a little farther south, could connect with the Niobrara River in Nebraska and move eastward, as evidenced by camera confirmation in 2004. In both cases, these cougars would eventually connect with the Missouri River upstream from Springfield, Iowa, and would share the same fate as animals leaving the Black Hills along the Cheyenne River. Many, it seems, continue straight south into the panhandle of Nebraska and to the North Platte River. There they join with any animals coming from Wyoming or Colorado, forming a crossroads of dispersal. This is evidenced by the high number of reported sightings and animals being killed in this region. Some propose that there is actually a sustaining population in this region, pointing to at least one incidence of a kitten being killed.

Animals moving east out of the panhandle along the North Platte quickly enter the sand hill country of Nebraska (see chap. 3, fig. 11). This region is approximately 50,000 to 60,000 square kilometers of grassland intermixed with various rivers, lakes, and marshland. The fact that few reported cougar sightings and corpses have come out of this region attests more to the probable security of the area for cougars rather than the lack of animals, at least dispersing ones. Discussions of whether the sand hills region could or is supporting a cougar population have been considered earlier. For now at least, it seems that for approximately two hundred kilometers along the north side of the river, cougars can move eastwardly with ease. On the south of the river there are higher levels of human activity and thus more chances of being detected, evident by the four cougars that were discovered and killed in Scotts Bluff, Nebraska. Though there are periodic areas of native vegetation and topographic relief along the southern shore of the North Platte River, it would be safer for cougars to stay on the north side.

About a third of the way across the state, the North Platte joins with the South Platte, which comes out of neighboring Colorado. Though the South Platte originates in the rugged mountains west of Fort Collins, Colorado, the fact that it passes through the heavily populated front range would discourage cougars from following the river to the east. If they did, they would find most of the riparian areas of the South Platte all the way to its junction with the

North Platte dominated by agricultural activity. Although possible, it is not very probable that cougars would move eastward along the South Platte, especially undetected. Shortly after the North and South Platte Rivers join, we lose the influence of the sand hill country, and the Platte River becomes surrounded on both sides by agricultural activity. However, the actual river valley itself is relatively wide (0.5–1.0 kilometer) over most of that way and contains fragmented but continuous tree or shrub cover. In its favor is the fact that most of the towns and cities located along it are small, and none completely encompass the river. Thus it is possible that if a cougar stayed alongside the river, it could move unnoticed along the Platte for an extensive distance. As mentioned in chapter 3, at least one cougar, documented by a trail camera in 2009, has done this. This cougar has not been seen in the area again, and it is assumed it moved on, supposedly to the east, where its fate remains unknown.

If cougars continue to follow the Platte, where will it bring them? Fortunately, the Platte River arrives to the Missouri just south of Omaha where their fates join with any cougars making it past this metropolitan area. Unfortunately, the riparian habitat along the Missouri to the south has been greatly reduced and fragmented, and what remains has little chance of supporting a population of cougars. However, from just south of Omaha to Kansas City, more than three hundred kilometers to the south, there is enough riparian and adjacent upland patches of forested habitat to allow a cougar traveling by night to successfully traverse the bare areas and spend the days in these patches. Additionally, there are only about a half dozen small towns along the way, with only half of them on the west side of the river. Any cougar coming down the Platte would most likely stay on the west side of the river, so most of these small towns could easily be bypassed. The largest city, St. Joseph, Missouri, is on the east side and could easily be avoided on the west bank where an oxbow in the river provides a narrow (about three kilometers) passageway between two small Kansas towns. Consequently, if a cougar traveling down the Platte makes it to the Missouri or if one somehow makes it past Omaha, he could move relatively undetected for about three hundred kilometers along the Missouri to the south. This trip comes to an abrupt halt upon reaching the twin Kansas Cities of Kansas and Missouri where the Kansas River joins the Missouri (fig. 16). There has been at least one sighting of a cougar, on the University of Kansas campus in Lawrence, so the possibility of cougars moving down the Republican and Blue Rivers to the Kansas River and then on to the Missouri exists. In any case, these cougars would also become linked with animals from the north as they move along the Missouri toward the Kansas Cities. Can they make it past this

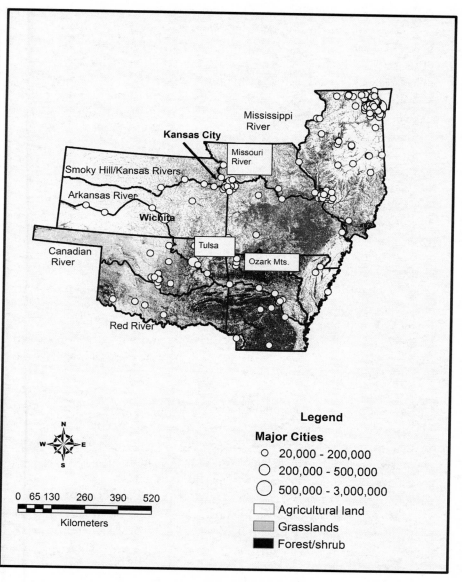

FIGURE 16. Patterns of forest, grassland, and agricultural land along major rivers running through southern plains states. Cities with more than twenty thousand inhabitants are indicated and those mentioned in the text are labeled. (Modified from data available from U.S. Geological Survey, Earth Resources Observation and Science [EROS] Center, Sioux Falls, SD.)

large metropolitan area? Animals coming in from the Kansas River would be hard-pressed to continue on this route, which leads right to the heart of the two cities. Their only hope would be to leave the river before arriving to the urban areas and then only if they leave to the north. To the south, they still run into a large wall of urban development extending thirty kilometers to the south to Olathe, Kansas. To the north there are some forested patches that they could use to connect with the Missouri north of the cities. For them and any cougars moving down the Missouri, the only hope to bypass the cities is to travel to the forested patches interspersed in this area. This is apparently what one young male tried to do in 2002 but was killed trying to cross Interstate 35. The fact that he made it to that point demonstrated that a cougar could travel unde-tected at least that far. If they can safely cross the many roads, then it is pos-sible that they could again join up with the Missouri River. From that point on, a wandering cougar has almost made it. There are no more large cities until St. Louis on the Mississippi River, and the number and sizes of forested patches increase as we near the Ozarks (see fig. 16). At any time after Kansas City, a cougar could turn south through farmland dotted with scattered forested patches and eventually reach the denser forest of the Ozarks region. In fact, after Jef-ferson City, Missouri, the landscape mosaic is dominated by forested habitat leading to the south. The confirmed sightings of cougars in northern Missouri since 2000 indicate cougars are making it to Missouri, probably by coming down the Missouri River.

In summary, the lower Missouri River provides a treacherous but passable corridor for cougars coming out of Nebraska and Kansas, and eventually mak-ing it to Missouri. Adding these movements to those out of North and South Dakota, we can see that the fate of cougars dispersing from a wide geographi-cal range is linked initially or ultimately to the Missouri River. As it was for the human pioneers moving west, the Missouri is an important conduit or cor-ridor for the modern-day cougar pioneers moving east. It appears that each year unknown numbers of young, mostly male, cougars start the journey or finish it along the banks of the Missouri River. How far they make it depends on the ecological health of the river. In more headwater regions where human popula-tions are low, the Missouri and its tributaries seem to be functioning well—at least for the dispersing cougar. The bottlenecks to movement for these ani-mals develop farther east along the agricultural wall that extends from North Dakota to southern Iowa. Although seasonal permeability seems to develop in this wall with the summer growing season, cougars arriving at the wall out-side of this season face the grim possibility of trying to move several hundred kilometers across a barren landscape. In this region, most rivers run in a southerly direction, and few river corridors exist to facilitate easterly movements of animals. Unless a national effort develops to reestablish native prairie patches in a stepping-stone corridor pattern across this region, successful movement of cougars through this region will be limited to midsummer and early fall. Farther south in more densely human populated areas, riparian habitat has been reduced, and large river cities present formidable bottlenecks to movement along both the Missouri and its tributaries. Although more degraded than upstream areas, these rivers still provide an avenue for easterly movement, an avenue that could be enhanced for the benefit of the river systems and for the cougars.

The health of our river systems is vital to the ecological functioning of the broader ecosystems in which the rivers are embedded. Because of their valuable ecological role, it should become a national goal to reestablish and protect riparian areas along all streams and rivers in the country. Each waterway should have a protected corridor surrounding it where riparian vegetation is revived. Because of the nature of riparian areas, this revival of riparian habitat in many areas can be as simple as retiring the land from agricultural use. Trees and shrubs in these areas would rapidly establish and convert open stretches of stream bank to shady riparian habitat reminiscent of pre-settlement times. The increase in shelter would benefit a variety of riparian species and provide cover for dispersing cougars along the Missouri and its tributaries; this would in turn reduce and in many areas eliminate bottlenecks to the dispersal of cougars across the plains area to the forested areas of northern Missouri. If we add to this a system of green belts around large river cities, besides enhancing the environmental esthetics of these cities, the forested cover of these green areas would provide sheltered passageway for the eastern migrating cougar. Thus, with some creative conservation practices that would enhance ecological integrity for our benefit, the travels of dispersing cougars would be imminently safer. Again, it all depends on the human will to facilitate these changes.

THE SOUTHERN ROUTES

There are at least four confirmed sightings of cougars in northern Missouri possibly attributable to movement along the Missouri corridor. There are various other sightings in southern Missouri, including some in 2011, as well as various sightings in Arkansas, all of which may indicate that other animals may be coming in from a more southern route. This turns our attention to the last three major west–east rivers crossing the plains: the Arkansas, the Canadian, and the

Red. Unlike the rivers we have talked about so far, they are not tributaries of the Missouri River but flow across the southern plains to eventually drain directly into the Mississippi. However, long before they reach the Mississippi, these three rivers enter the forested areas of eastern Oklahoma and then the Ozarks of Arkansas. Once a cougar reaches these areas, it is basically home free. Eastern Oklahoma has sufficient forested habitat to support cougars and that same forested habitat will provide ample cover for cougars to reach the Ozarks, which we have also identified earlier as an area of ample habitat, of both vegetation and topography, to biologically support a sizable cougar population. The question then becomes: Can they make it from the mountainous areas of eastern New Mexico and Colorado across the open plains to the forested eastern regions? Based on the analysis of LaRue and Nielson, their chances seem to be pretty good, and it is this southern route that LaRue and Nielson identified as the most feasible pathway for cougar movement to the east.

The first river to consider is the Arkansas River. The Arkansas is the most northerly river that is not a tributary of the Missouri and is the one that reaches the deepest into the Colorado Rocky Mountains where healthy cougar populations exist. Its effective reach into cougar territory is amplified by the many tributaries that flow out of the multitude of mountains and canyons of south-central Colorado. So, unlike all the rivers considered so far, the Arkansas taps into a substantially larger source population of cougars. Dispersing animals from an area in excess of 30,000 square kilometers can potentially begin their journey down the Arkansas or one of its tributaries. Thus, the number of dispersers heading east alone makes this route the most feasible for pioneering cougars. Added to this potential high number of dispersers is the fact that the first four hundred or more kilometers of the journey are either in the mountainous habitat of Colorado or through the front range of Colorado with the lowest human density of the region. The only major city is Pueblo, Colorado (pop. 105,000), located in the foothills of the mountains. The river runs right through the city, and a city of this size in the plains states might present a formidable barrier to moving cougars. Just upstream from Pueblo is the Pueblo Reservoir on the river, and any cougar coming down the Arkansas upon reaching the reservoir would probably leave the river course at the dam site, especially on seeing the urban area below. Fortunately, there is a lot of native vegetation surrounding Pueblo, and circumventing the city either to the south or north would be relatively easy. In doing so, a cougar would run into the many tributaries leading back to the main river downstream from Pueblo. Also, most of the tributaries coming from the southern mountainous areas enter the Arkansas to the east of Pueblo, thus bypassing the city. From Pueblo eastward in Colorado there is some agricultural activity along the river. It is not very extensive, though, and is embedded in a landscape of native vegetation, which a cougar could easily use to bypass the cultivated lands. As a result, cougars coming out of the southeastern Colorado Rockies would find few barriers to their movements along the Arkansas River up to within thirty kilometers of the Kansas border. At that point the Arkansas River enters a region of intensive agriculture where native vegetation in turn becomes embedded in an agricultural landscape (see fig. 16). From here to the Cross Timber areas of Oklahoma (see chap. 3), the ability of a cougar to move along the river will depend on how much riparian vegetation still occurs. Fortunately, for most of the four hundred kilometers of river from the border to Hutchinson, Kansas, there are only a few small towns along the river. Most of these towns do not encompass the river but are located off to one side or the other. There is still ample wooded habitat along most of the river, but where there isn't, the gaps are narrow enough that a cougar could easily pass through during the night. There are even a few areas of native vegetation, including trees, which are extensive enough that a cougar could easily pass a few days resting with little chance of being discovered. One exception is Dodge City, located about halfway along the route. Dodge City (pop. 25,000) is relatively small, but the river runs right through it and has little riparian vegetation. However, the distance from the west to the east side of the city is only about three kilometers, and a cougar would be able to pass through undetected during the night. This is also true for the city of Hutchinson (pop. 41,000) where the river passes along its southern edge (South Hutchinson), and in this case, there are some small patches of forested habitat for cover. Thus, a cougar moving down the Arkansas River from Colorado would have a long but hopefully uneventful trip with regards to contact with humans until it comes to Wichita (pop. 345,000).

Just as the Arkansas River enters Wichita from the north, it has been channelized and much of the water diverted to a scenic waterway that runs through the heart of the city. The original channel has little or no vegetation, and the diverted streamway is confined in a cement channel. For a cougar moving down along the Arkansas River, this landscape, together with Wichita's size, represents a migratory dead end. The only hope a cougar has of bypassing Wichita is to leave the river valley just after Hutchinson, travel south over farmland for about fifteen kilometers, and join up with the Ninnescah River just above the Cheney Reservoir. There is little native vegetation in this area but also few houses. The main field crops grown in this area are wheat and corn and so, farther to the

north and during the right time of the year, the farm crops could provide adequate cover to make the journey. Once connecting with the Ninnescah River,
the cougar could follow it to the southeast where the Ninnescah and others
join with the Arkansas River south of Wichita. From that point on there are
few towns and again sufficient riparian vegetation for cover up to the Oklahoma border. Shortly before reaching Oklahoma, the Arkansas River borders
the Flint Hills region of Kansas. This region, along with the Osage tribal lands
in Oklahoma, could biologically support a small cougar population and provide adequate shelter for dispersing animals. Once crossing into Oklahoma, a
cougar would enter the Cross Timber country where there is adequate habitat
for dispersing individuals. Cougars would, however, have to leave the Arkansas
River valley somewhere before it reaches Tulsa, Oklahoma, which would provide
a significant barrier for further river travel. Fortunately, as mentioned, the Cross
Timber area, especially south of Tulsa, provides ample forest cover to stay in or
continue moving to the east to the Ozark and Ouachita forest areas.

One last factor to consider regarding cougars moving down the Arkansas
River is the possibility that they might leave this river system early on in their
journey. Specifically, this may have happened just east of Dodge City. After
Dodge City, the Arkansas River heads in a southeasterly direction for about
thirty kilometers. At that point it abruptly turns northeast. Just twenty to thirty
kilometers south of that bend in the river are the headwaters of Cimarron,
Medicine Lodge, and Salt Rivers, all tributaries of the Arkansas that join it next
to the Osage tribal lands. By continuing south at this point and entering the
course-way of any of these rivers, a cougar would reduce its travel significantly.
It would also quickly enter river systems, which provide more cover and fewer
humans than staying the course along the Arkansas. This would all facilitate its
passage to the Cross Timbers region and forested lands beyond.

Why would a cougar leave the Arkansas River at this point if he has no prior
knowledge of these waterways to the south? One reason is that he is heading
in a southeasterly direction already, and instead of taking the abrupt turn to
the north, he could as easily continue on his current heading. There are a few
small tributaries of the Arkansas, which come in from the south in this area,
that would provide cover to entice a cougar to continue heading that way. Again,
at certain times of the year, wheat and corn crops provide additional cover. If
our cougars do head that way, in a very short time they would come upon the
fairly extensive areas of natural vegetation, including some forest patches. Would
they do it? The one cougar that was killed in the Medicine Lodge watershed in
2007 suggests that at least one animal may have continued heading southeast off

of the Arkansas at that point. Enhancing the possibility of cougars continuing southeast here would improve the feasibility of the Arkansas and its tributaries as a successful dispersal corridor. It would also reduce the possibility that cougars would continue along the Arkansas where they would be faced with the eventual barrier presented by Wichita.

The next river to consider is the Canadian. Like the Arkansas, the Canadian River has its headwaters in the Rocky Mountains near the border of New Mexico with Colorado. This river and its tributaries drain current cougar range and so provide a convenient exit for dispersing animals. From its headwaters, the Canadian runs for more than 1,000 kilometers out of New Mexico, across the panhandle of Texas, and into central Oklahoma where it eventually joins with the Arkansas near the Ouachita Mountains. From its beginning to western Oklahoma, the Canadian passes through some of the least populated landscape in the region. There are few towns and cities, and very little agricultural activity. The river passes by Amarillo, Texas (pop. 174,000), but it is more than twenty kilometers from the city's northern limits, with little human presence. There are scattered forest patches along the river where a dispersing cougar could rest and, as they did historically, find deer to hunt. The level of agricultural activity increases as it enters Oklahoma, but so does the amount of forest cover. In fact, it is this area in Oklahoma, the western extreme of the Cross Timbers region, that could possibly support a small population of cougars. Though it is uncertain whether the area could support cougars, it is more a certainty that they could at least pass through the area with little problem. Farther to the east, the Canadian River passes close to Oklahoma City and then Norman. Under different circumstances, the proximity of the river to this large metropolitan area (pop. greater than 1.2 million) might pose a formidable barrier to cougar movement. However, because of the scattered but abundant forest patches and the low level of intensive agriculture, any cougar can skirt the cities, mainly to the south, and enter into the Cross Timbers (see chap. 3, fig. 12) region of the state and on to the Ouachita and Ozark mountains.

Because of the link to a source population and the ease at which a cougar could travel along most of the river, the Canadian River is most likely one of the specific travel paths within the zone that LaRue and Nielson identified as a dispersal corridor. Another specific travel path is the Red River of the South.

This Red River forms much of the border between Oklahoma and Texas. It is not as long as the Canadian River, with its headwaters in the panhandle of Texas just south of Amarillo. Although the Red River does not directly connect with the mountains to the west, any cougars coming out of southern New

Mexico or western Texas, specifically along the Pecos River, would have a high probability of joining up with the Red River. Once along this river, again, there are few towns; where there is agricultural activity, mainly winter wheat, there is sufficient native vegetation along the river for cougars to pass by relatively undetected. About halfway across the state of Oklahoma, the Cross Timbers country begins, providing additional cover. Farther to the east, the Red River passes about eighty kilometers to the north of the Dallas–Fort Worth area (pop. greater than 6.4 million). Fortunately, this distance, along with the low agricultural activity and the occurrence of forest patches near the river, mitigates what could have been a major barrier to cougar movement. Once passing Dallas–Fort Worth, the dispersing cougar would easily enter the Ouachita Mountains region.

In summary, it is easy to see why the least-costs path analysis of LaRue and Nielson selected the gateway between Oklahoma City and Dallas–Fort Worth as the most likely corridor for eastern movements of cougars. The cities are separated by more than 250 kilometers in which there are no large cities and there is ample forest cover for a traveling cat. In their analysis, however, they indicate the most likely source of cougars using this corridor would be from southwest Texas. Although it is very likely that cats from this region of Texas could move to the northeast to the passage, they would have to move more than 400 kilometers, crossing river systems rather than moving along them. Given the propensity of cougars to follow rivers, especially in open landscapes and not knowing that the northeast passageway exists, most dispersing cougars from southwest Texas would more likely move either to the north along the Pecos River or to the east along the Colorado River of Texas or the many other smaller rivers that flow to the east. Whether these eastern-moving cats could make it past Austin and San Antonio is uncertain. The fact that there are reports of cougars from southeastern Texas and Louisiana indicate that animals may be dispersing along these southern rivers to the east. Cougars moving up the Pecos River could eventually link up with the Red River and then move east.

An additional factor to consider is this: cougars are considered varmints in Texas and can be hunted year round. This means the number of possible dispersing animals from Texas would be lower than from eastern New Mexico where there is a regulated season and thus possibly higher numbers of dispersal-age animals. Given these factors, I believe that most dispersing animals using the Oklahoma City/Dallas–Fort Worth passage would originate from New Mexico. As animals become established to the east and DNA comparisons can be made, the origin of the founders can be determined and the two different predictions tested. In either case, once reaching this passage zone, cougars could probably move eastward through most of the area between Oklahoma City and Dallas–Fort Worth. Because of their tendency to follow rivers, however, their main pathways would be along the Canadian and the Red Rivers. Although no major obstacles appear to exist at this time to the movement along those two rivers, a good conservation strategy would be to identify potential barriers and take appropriate action to ensure future connectivity along the rivers.

Based on my visual analyses, I have proposed various different possible dispersal routes for cougars to move across the plains/farmland and eventually into the eastern forests. In some cases these predicted corridors correspond well to what LaRue and Nielson proposed based on an objective least-cost pathway analysis. The least-cost pathway analysis provides us with a good overview of possible general travel routes. How well it performs depends on the factors entered into the models and weights given those factors relative to their facilitating or encumbering cougar travel. The purpose of my analysis was to go beyond the general patterns and use what we know about the behavior of cougars to see if we could identify specific travel pathways. In doing so, I have attempted to take virtual dispersal journeys across the plains, going and turning where I thought a cougar would in order to remain hidden and find food. That some animals may have taken these same pathways is verified by the confirmed sightings of cougars at the end points. Whether those animals took the exact paths that I did is unknown. It is my best estimate to try to give us specifics regarding potential movements of cougars across an often hostile landscape. This will allow us to identify where we might try to document these movements and also ensure safe passage where there are bottlenecks or barriers to cougar movement. In some cases, for example rivers running through major cities, these bottlenecks cannot be avoided, and any cougar taking these paths will most likely be discovered. If possible, we would like to discourage cougars from starting down the long journey that will end in death or a life in a zoo. In other cases, we can enhance the permeability of these barriers, for example by planting shrubs, setting aside riparian habitat, or planting specific crops. In doing so, we may facilitate the safe movement of these intrepid feline pioneers as they attempt to find a new home and life in the East.

Why do we want to help these dispersing animals make this journey? The answer is long and complex but suffice it to say that forested ecosystems of the eastern third of the United States have suffered too long without the presence of their top predators, in this case wolves and cougars. An ecosystem without its apex predators is like a spruce tree with its top cut off. Unable to continue to reach skyward with its terminal branch, the tree's biomass becomes concentrated in its lateral branches, each fighting for dominance, none winning, spreading farther and farther out sideways. The tree survives but it is deformed, dysfunctional. And this seems to have been the case with the eastern forests. Ecosystems function through energy flowing into and eventually out of their components. Deprived of top predators, this energy, rather than flowing up to the top, stays concentrated in the lower prey levels, in excess numbers of prey, in this case deer. This blockage of energy and these excess deer have begun to deform the ecosystem and make it dysfunctional. The deer strip the forest of their preferred plant species, reducing overall plant diversity. They enable unpalatable species, natives and exotics, to flourish, reducing forest floors to monocultures devoid of places for nests and burrows of other wildlife species. This has been happening in the eastern forests for more than one hundred years. It is time to rescue these great ecosystems from this growth of uncontrolled prey. The simplest and fastest solution to the problem would be to simply reintroduce cougars, and for that matter wolves, back into these areas. However, in this present-day atmosphere of culpability, we seem to be reluctant and incapable as a society to take bold conservation actions. Public officials fear for their careers more than for the health of our ecosystems they are charged with protecting. Hunter conservationists of old have been replaced by modern-day hunters who view the hunting experience as a competitive sport, to "win" at any cost, in this case at the expense of the ecosystem. Conservationists have long been ridiculed and marginalized by moneyed interests or by those who concentrate on the more exotic and less controversial overseas species. Thus, reintroduction of cougars to the east is not likely in the near future, and our forests continue to suffer. The second solution is to bypass all of this inability to act and enhance the possibility that cougars will arrive on their own and free us all of any responsibility. Because of the recent and increased sightings of cougars on the eastern edge of the prairies, there is hope that this species will do what we lack the will to do. If cougars can reestablish themselves in the Ozarks, in the northern forests of Minnesota and Wisconsin, they will be poised to make the final push eastward. Like those first Europeans colonizing the eastern shoreline, they can spread across the vast vacant forests, fulfilling their manifest destiny to reclaim their ancestral lands. Upon doing so, the healing of the eastern forests will begin; the righting of a terrible ecological wrong will be accomplished. I wish these brave pioneers all the luck in the world.

5 Challenges Facing the New Pioneers

COUGARS, AFTER A HIATUS OF MORE THAN A CENTURY, are starting to move into and, in some cases, re-colonize parts of the Midwest and Great Plains. I have alluded to where this is happening and used some of the confirmed sightings of cougars in my discussions of where they may be able to reestablish populations (chap. 3) or which routes they may take while moving eastward across the plains (chap. 4). As mentioned, every state I have considered so far has had confirmed sightings of cougars, some more than others. Before 2000, there were documented sightings in many of these states, but most, and increasingly more, of the sightings have been after the turn of the century. This is indeed the century of the cougar. It is tempting to present a detailed listing of all the current confirmed or at least reliable sightings, and many have done so for specific states. However, things are happening so fast regarding cougars in the prairie region that any specifics I give would quickly become outdated, supplanted by even newer data. Instead, I can refer you to the website of the Cougar Network (http://www.cougarnet.org) for the current (and future) state of affairs regarding cougar sightings. From that site and from the increasing number of newspaper articles on sightings, one can stay informed and updated as to what is happening regarding cougars in the region.

Though I cannot give you any specifics, I refer you back to chapter 3 regarding my predictions as to where cougars will reestablish populations. I also refer you to the 2011 article by LaRue and Nielsen. Based on those predictions, one can see that eventually every state in the prairie region will have to contend not only with animals moving through their borders but also with established breeding populations. Some states will have larger populations than others, but all will be faced with the challenges of cougars rejoining the list of native fauna. Currently, South and North Dakota are the only two states to have established populations of cougars. Additionally, Nebraska is very close to acknowledging the presence of a breeding population in its western region. These states are faced with the immediate reality of cougars, while others are dealing with the anticipation of having them in the not too distant future. Cougars will return, and the question then becomes: How does a state deal with the return of a top predator to its boundaries? The response to that question lies in the answer to how possible it is that we will allow cougars to return. Throughout this book I have referred to the biological and political/sociological ability of an area to support cougars, primarily concentrating on the biological. Now it is time to turn our full attention to the more political/sociological aspects of cougar re-colonization of the Midwest/Plains. These aspects center on how accepting people are of cougars in their midst and what adjustments we are willing to make to accommodate them. Ultimately, it is the level of this acceptance that will determine whether viable populations of cougars can return to the many parts of the region I have identified.

Given that there are areas in the prairie region that can biologically support cougars, what would they face if/when they do return? This consideration is important because the social and political realities of cougars in the prairies could be the deciding factor on the fate of these dispersing animals. Because people, a lot of people, live with cougars in the West, we may not foresee major problems with people in the plains adjusting to the arrival of cougars. We can argue that if westerners are courageous enough to peacefully coexist with cougars, would their midwestern relatives, also of pioneer stock, be equally intrepid? Would they show the same level of tolerance, of acceptance? Surely, if asked, they would proudly claim they are as fiercely independent and resilient as their more western counterparts. But are they, especially when it comes to cougars?

Unfortunately, unlike westerners who were born and raised with cougars on the landscape, people in the plains region have lived without cougars for more than one hundred years. These residents of the Midwest may still profess to have that pioneer spirit, but it was their pioneer ancestors who lived with cougars. Though natives to the region, cougars are new on the scene and, for most current human inhabitants of the region, an unknown and scary factor. Cougars are a large predator, known to have killed some humans, fewer by far than the family dog, but still, these are wild animals. We have come to expect that wild animals in the Midwest do not normally attack humans.

Unlike in the West where cougars currently exist, the prairie region does not have mountainous areas that function as refuges for wildlife in general and cougars in particular. There are no large national parks and fewer federal lands where the presence of large predators is accepted or tolerated. In the West, most people are concentrated in towns and cities, located in treeless valleys, away from the areas preferred by cougars. The result is that in many areas, there are clear boundaries between where people and wildlife have reciprocal priority. When we go to the mountains, we expect to make some compromises, to give wild animals the temporary respect they deserve. In the Midwest, those boundaries are not as clear. Though overall human density of midwestern states is similar to that of western states (3.6 to 20.7 vs. 2.5 to 90.5 persons per square kilometer, respectively), people and their presence (farmlands, homes) are more evenly spread out in the prairies. Most areas of the Midwest are used mutually by humans and wildlife, putting them into more direct contact and conflicts. Deer, for example, are found in their highest densities in the farmlands and suburbs of the Midwest. As a result, for cougars to coexist with humans in the Midwest, cougars and humans need to confront the possible greater contact they will have with each other than they currently do in more western states. Politics and social acceptance become more the deciding factor regarding cougars' survival in an area. These two factors—the unfamiliarity of people with cougars and the difference in the ways humans and cougars might position themselves on the landscape—can present many real or imagined social and political problems with the return of cougars.

As a result of all this, the success or failure of the return of cougars to the prairie landscape depends on their acceptance or rejection by the humans living there. In considering this we have to first realize that the lack of acceptance is not because the cougar is a predator or because it kills deer. Midwestern people live with, or at least tolerate, a variety of predators from weasels to coyotes, including a smaller wildcat, the bobcat. What will affect peoples' political and social acceptance of cougars is the danger or lethality, real or perceived, of this predator to humans and, to a lesser extent, their domestic animals. Because of the spatial separation of humans and cougars in the West, their real and perceived risk is generally low. Or, at least it is temporary, only when people enter "cougar country." Consequently, in this chapter the first thing to address is: Just how dangerous is a cougar? Once we have an idea of what risks people in the Midwest would face with cougars, I can attempt to address what type of social and political environment cougars should face when moving into these uncharted areas and compare/contrast that with what they are facing in some of the areas where they have already appeared. From these discussions we can get a better feeling of whether and where cougars might be able to survive and what adjustments humans may or may not need to make to ensure the safe return of cougars for all concerned.

This may have been the hardest chapter for me to write for several reasons. Because there are various studies and hard data on the lethality of cougars, the first part was easy. The hard part came from attempting to analyze humans' reactions to cougars over eleven different political and cultural entities. Each state has its own history, its own direction of development. Some are more rural than others, some more urban. Some have large relatively undisturbed areas, more forested areas; others have some of the highest concentration of agricultural activity in the world. Each can and will view the return of cougars with their particular perspectives, biases, and concerns. A second reason is that there is so much recent and new, up-to-date information on how citizens and their governments are reacting. Reactions can change as a state goes from potentially having cougars to actually having a breeding population, as I explain later. This makes it hard to generalize on how cougars are currently being accepted/rejected.

The main reason this has been difficult, however, is because I have had to try and separate what I as a wildlife ecologist, who sees the value of top predators, think the reactions of people facing this new challenge should be, and how I as a citizen and a father would react to the arrival of a large predator to my state. In trying to separate these and other emotions I and others have concerning this topic, I have attempted to best present the facts and address the concerns in order to present a good objective data base for readers to form their own opinions. This is the goal of this chapter: to inform more than to convince; regardless of what decision a town, a city, a county, or a state makes regarding cougars in their midst, it should be based on rational consideration of the facts as we know them. I do not expect those decisions to coincide with what I would decide, but whatever they are, they need to be defendable by the knowledge they were based on, not on cultural baggage, fear, ignorance, or political gain.

How Dangerous Are Cougars to Humans?

All sources indicate that there has yet to be an authenticated report of a healthy wolf killing a human in North America, and attacks are rare. Though some may dispute those claims for wolves, no such dispute occurs relative to cougars. There is no doubt that cougars, presumably healthy ones, have killed people in the past and have done so even fairly recently, but does that mean they are dangerous? Should we be afraid of them, of their return to our state? Maybe they are just too risky to have around people under any circumstances. If that is the case, then is this the end of the discussion and the end of the chapter? No cougars, no way, no how?

As you would suspect, that is not exactly the case. Cougars have in fact killed people, but we know that a lot of things, including other people, kill humans, lots of humans. Do we consider these sources of death dangerous? Are they dangerous enough to be concerned, to alter our behavior, to be afraid? The fact we could be killed by *something* is commonly accepted in society. Daily, we put ourselves in risk of being killed by a myriad of really dangerous things: cars, random shootings, accidents, and the like. However, most of us willingly take that risk. How do we decide, consciously or subconsciously, if exposing ourselves to something that could kill us is worth the risk? When does a potentially deadly agent or action become dangerous enough to do something about it?

Whether we consider something as dangerous or as an unacceptable risk depends on just how often it occurs and thus what is the actual risk or probability of it happening to us. The whole science of risk assessment, which is the backbone of the insurance business, centers around calculating risks—risk of injury, risk of economic loss, risk of death—from a whole array of causes. Upon knowing the risk of a certain action, insurance companies set their rates. We, as individuals, make decisions, consciously or not, to take that risk. This is as it should be with cougars. Like all the other risk factors in our lives, we need to first know the likelihood that cougars will actually affect us, and then use that information in making our decisions as to how afraid we should be of them, whether we want to "take the cougar risk."

The first step is to know just how many people cougars have killed and over what time frame. Paul Beier, another well-known cougar biologist, did just that. He examined all the records he could find regarding attacks on, and deaths of, humans by cougars in the United States and Canada from 1890 to 1990.[1] Over that one-hundred-year period, he was able to verify nine attacks resulting in ten people being killed and forty-four nonfatal attacks resulting in forty-eight nonfatal injuries. Others have analyzed the more recent records and from 1991 through 2010 report there have been eleven deaths attributed to cougars and eighty-nine nonfatal attacks reported over that time period. Although the sources for the reporting of these later attacks have not been put under the same scrutiny as those reported by Beier, most of the fatalities do, in fact, seem attributable to cougars. Whether these more recent numbers represent an increase in the number of attacks and fatalities over the previous one hundred years, or just increased news coverage and better medical reporting, is unknown.

There easily could have been incidents of cougar–human interactions in the late 1800s and early 1900s that have gone unreported. Much of America was a wilder place back then, especially where cougars still roamed.

Although we don't know the accuracy of the reports, if we take the later estimates for 1991–2010, it appears that we could expect an average of 5 attacks and 0.55 fatalities per year from cougars on people in current cougar range. The questions then become: Is that high? Is it an unacceptable risk? Should we be afraid of cougars? Here is where the risk assessment comes in. To decide if the chances of being attacked or killed by a cougar are high, we need to compare them to other risks we take by examining how many people get killed or injured (attacked) from different causes to put the risk of a cougar attacking or killing you in perspective.

To start with, how dangerous are cougars compared to other causes of death and injury we potentially face? I leave out the "big killers" such as the risk we take smoking (467,000 deaths per year) or overeating (216,000 deaths per year), mainly because they are self-inflicted. It is worth noting that although these two causes outpace all others, if we are overweight and go for a smoke in cougar country, we are likely to be more afraid we will die from a cougar attack. Here I concentrate on injury or death caused by "accidents," first from human-related causes and then specifically those causes related to other wildlife species.

After setting aside most of the self-inflicted causes of injury or death, let's start off with the most dangerous one, driving a car. Unlike health-related deaths, even healthy, cautious drivers can get into accidents and get killed. How does that risk compare to being attacked or killed by a cougar? Driving and riding in a car is probably one of the most dangerous activities we can engage in. Annually 40,000 to 50,000 people are killed in car accidents, and millions are injured. Based on these numbers, each day we get into a car we literally place ourselves in mortal danger. Yet we take the risk daily. Even walking on the streets is more dangerous than walking in the woods: 5,000 persons per year die from being hit by cars and another 65,000 per year are injured.[2] More people die per year from riding bicycles (759) than would be killed by cougars in a thousand years. Enough of us play with dangerous fireworks to be killed per year equal to the total number of humans killed in the last twenty years by cougars. We shoot each other, accidently, while hunting at a rate of over five hundred per year, fifty-five of them fatally. And it is not only outside in the wild; statistics tell us just staying in the house is dangerous. Toilets cause more than 43,000 injures per year. In fact, the home is consistently listed as one of the most dangerous places to be. Every year hundreds of thousands of us die from a multitude of causes. Yet we take these risks, and more, every day of our lives. We enter our cars, we ride our bikes, we go hunting with friends. If we are willing to take these risks, why should the miniscule risk of a cougar attack keep us out of the woods? Driving to the woods to walk is eminently more dangerous than the walk itself.

Point made, but these are all risks from our stuff, our actions. Admittedly, life in the modern world is dangerous, but do we have a choice? How many of us want to get rid of toilet seats or cars? Is it fair to compare the risk of being attacked or killed by a cougar to these human-created dangers? How about comparing the risk of cougar attacks to those from other animals? Would that not be a better comparison to judge how dangerous cougars are from similar sources? From our earlier discussion, we know that cougars are more dangerous than wolves. With an average number of fatalities of 0.55 per year, they are slightly more dangerous than sharks (0.4 fatalities per year). So yes, cougars are more dangerous than some animals, but the list is short. Black bears, Yogi and Boo-Boo, Smoky, and the gang are more than twice as dangerous as cougars, killing 1.5 people per year. Even snakes kill more of us (15) yearly than cougars. So among some of the better-known wild predators, cougars are toward the bottom, along with the other two species, wolves and sharks, we seem to fear the most. Cougars are also far less dangerous than man's best friend, with dogs killing twenty-five to thirty people per year, many of them children. Yet we let these dangerous animals live within our homes.

Even though dogs are more dangerous than cougars, they are not at the top of the list of animals that kill people. There is one that is easily ten times more dangerous than dogs, killing as many as 200 people a year. Who is this vicious beast, this wild animal that is responsible for such savagery? Which fierce predator can it be? The grizzly bear? The polar bear? Maybe it is the wily but deadly coyote? All wrong answers. The deadliest of wild and domestic vertebrate species is not a ferocious predator; it is none other than good old Bambi. Yes, big-eyed, floppy-eared deer are the most dangerous species, not only killing hundreds of people a year but injuring (attacking!) an additional 13,000 people per year.

How do they do it? We don't often hear accounts of deer sneaking up and pouncing on unsuspecting humans, though it has been reported.[3] No assessment of the number of deer directly attacking or killing people could be found, but judging from the number of news accounts, deer may actually attack and kill more often than cougars. For example, in 2005–6 on the Southern Illinois–Carbondale campus, there were thirteen people injured by deer attacks.[4]

Deer may or may not be more aggressive and dangerous than cougars regarding direct attacks on humans, but this is not how they are the deadliest. So what is their preferred modus operandi, their favored method of attack? Deer wreak their greatest havoc on us while we are engaged in that other extremely dangerous activity discussed before, driving our cars. Each year there are over 1.5 million deer–automobile accidents in the United States, leading to those hundreds of deaths and thousands of injuries. From simple fender-benders to total wrecks, deer scare, scar, or kill more people in one week than cougars have over the last 120 years. Yet we fear cougars more. While heeding a warning not to go on a trail where a cougar was sighted, we jump in our cars and drive deer-infested roadways at dawn and dusk, when the chance of an "attack" are the greatest. We devour the once- or twice-annual news report of a faraway attack of a human by a cougar while skimming over the local news of yet another person being killed by a deer. Even when deer directly attack humans, as noted earlier, they are excused because males were "in the rut" or females were protecting their young. Additionally, people being attacked are often portrayed as being less than manly, rather than hapless victims of a vicious, dangerous animal, as they are when attacked by a cougar.

If deer are so much more deadly than cougars, why do we fear cougars more than Bambi? Why does the rare attack of a person by a cougar make headline news or go viral on the Internet while attacks by deer rarely make the news at all? Surely the lives lost from deer are just as meaningful, especially to the victims and families, as those few lost to cougars. Do we mourn the loss of our loved ones less if they were killed by a deer? It appears that as a society we do. We dwell on eleven unfortunate individuals killed by cougars in the last twenty years, know them by name, their life history, and the gruesome details of the incident. Yet we hardly give notice to the approximately 4,000 unfortunate humans killed by deer in the same period. Are not the hundreds of lives of children lost to deer as tragic as the four lost to cougars in the last twenty years? Was not the promise of their future lives, snuffed out all too early, as heartbreaking as the others? Simply by our greater media exposure to the deaths associated with cougars, we as a society demonstrate that we value those lives lost more. Why is that so? Is being attacked or killed by a cougar more "gruesome" or disfiguring? I argue not; many of the injuries humans sustain from these deer attacks are more gruesome and disfiguring than even the most severe cougar assaults. In many of the deadly "attacks," deer and humans are reduced to lifeless, mutilated piles of flesh more grisly than the fatal victims of cougars. So it can't be the gruesome nature of cougar attacks.

Is it because we feel that the deer didn't really mean to do this, did not purposely throw themselves at the car? Is it somehow not their fault? Evidence suggests that in most cases of cougar attacks, cougars didn't mean to do it either; often it is mistaken identity of a human for natural prey. You may find this surprising, but cougars don't really hate us. They do not purposely stalk a human but rather what they think is prey. Many of the attacks are of people running or jogging in cougar habitat. As deer get excited and focused on escaping just before colliding with a car, so too does the cougar; sighting a running object and with little time to determine if it is food or not, a cougar becomes excited and focused on the chase. In both cases, accidents happen. In both cases, there are terrifying moments for the people involved seconds before the deer or the cougar strikes. In both cases, for the injured, for the dead, it really does not matter who inflicted the wound or why, who took away that life. The pain is not less, the scars are not any less disfiguring, and dead is dead.

Why, then, do we publicize, dramatize, the rare occurrences of cougar attacks and even rarer cases of being killed by them? I suspect the main reason is because cougar attacks are so rare. And because of that rarity, when they do happen, it is news. With around thirty-five deer–car collisions per day and around five mortalities per week, though tragic, they are just not newsworthy. We are a news-oriented society. Each day we are bombarded by reports on the multitude of events occurring locally, globally. Because of this inundation of information and our relatively short attention spans, our minds seek the new, the different. Viewers would soon tire of hearing about another person, even a child, being killed in a deer–auto crash. It is just not "news" enough. These instances may make the local reports, if they are gruesome enough, but more than likely are only entered in the police log buried deep in the newspaper. Rarely does notification of these accidents and fatalities extend beyond state lines. What happens in Colorado stays in Colorado. Yet a human being attacked by a large, wild animal is such a rare event that the whole nation hears about it.

What this means is that the cougar is receiving all this negative press simply because it is such a good neighbor. Given the number of cougars and their often close association with humans, rarely are there problems. There have been between forty and one hundred panthers in southern Florida for eons. Even now with the massive human development in the region, there has not been a single panther attack recorded on humans. This is in contrast to more than ten panthers killed every year on highways in that region. Indeed, panthers have more to fear from us. Similar patterns of safety can be documented over most of the West where cougars occur. Though we can identify at least one hundred places in the United States and Canada where cougar attacks have occurred in the last 20 years, this means there are thousands of places where they have never occurred over the last 120 years. Yellowstone Park, with its tens of thousands of visitors per year, has not had a single cougar incident reported in the last thirty years. In contrast, bison attacked seventy-nine people over a twenty-year span (1980–99).[5] In Jackson Hole, Wyoming, a town embedded in cougar habitat, there is not an incident on record. And this is in light of the fact that a female cougar raised her litter in plain view of the town. The list goes on and on; there have been thousands of families living in the mountains for generations who have coexisted with cougars without incident. As the statistics show, cougars are not that dangerous in their current range and that is not expected to be different when they move to the Midwest. Most recent studies consistently demonstrate that cougars really do go out of their way to avoid human contact.

That said, can we assume that for any given area, an incident with cougars will never occur? The chances of that occurring in any given place are infinitesimally small, especially with regards to all the other things out there in rural America that could injure or kill us. In 2008, 456 people died in farm-related accidents, more than a hundred of them children.[6] Not that being attacked or killed by a cougar is not tragic; it is just as unfortunate as being killed by a car, a tractor, or a bee, or any other thing. It is just that, statistically, it is not likely to happen to you or your children. So, yes, if cougars return to the Midwest, there may be incidents. The chances of those are a lot less than from other animals, including those two icons of tranquility, dogs and deer. The probability of being attacked by a cougar does not justify the hysteria and fear that often surrounds any discussion of cougars re-colonizing the prairie states. These facts about the risk of cougars should ring loud and clear in these debates rather than the perceived myths and unfounded beliefs. Later, we will see if that is the case.

How about cougars attacking other animals, especially domestic livestock and wildlife of interest such as deer and elk? Even though we may not need to fear for ourselves upon the return of cougars, would they inflict unbearable economic loss on local ranching operations or decimate our deer herds? These concerns for both domestic and wild "livestock" can also be major detriments to cougars being welcomed back to their former hunting grounds. So the question now becomes: Just how dangerous are cougars to domestic livestock and wild ungulates? To answer that, we need to turn our attention back to the West and the Southwest where cougars coexist with livestock and wild ungulates.

First, considering livestock, there is no doubt that cougars kill domestic livestock. How many and how often seems to depend on the type of livestock, and where and how they are grown. Regarding the type of livestock, we can divide livestock that might be impacted by cougars into large and medium sized. Large livestock includes cows and horses while medium-sized ones include primarily sheep and goats. I do not include smaller species such as chickens because cougars so rarely attack these animals.

Adult cows, weighing about a half ton, have little to fear from cougars. Reports document occasional predation of adult cows by cougars. However, they are few and far between. In the area in southern Idaho and northwest Utah where I studied cougars for sixteen years, there were both cougars and cattle in the area. Over that time period, I did not find one incident nor did I hear of any report of an adult cow being killed by a cougar. In contrast, I did find one incident where a small tree, a mountain mahogany, had killed a cow. Apparently this animal stuck its head through a V formed by two low branches and could not escape. She eventually died from the stress of the struggle and/or starvation. So the statistics from my study area indicate trees are more dangerous to cows than cougars.

So much for statistics. There are reports from elsewhere of cougars killing adult cows, but very few. Why is that so? One would think a large source of meat such as a cow would be very attractive to a cougar. Why not kill and eat them? Within the answer to that question lies the dilemma for all predators, the cost-benefit ratio of prey. On the benefits side, a cow presents a large food source. However, on the cost side, a cow is a formidable adversary, making it difficult and dangerous to try and attack it. The difficulty for the cougar lies in the fact that the preferred point of attack is the neck region of his prey. When attacking, a cougar strives to jump on the back of the animal and either give a lethal bite to the neck or try and pull the animal's head back and break the neck. When it comes to cows, their necks are too big to pierce with the relatively short canines of the cougar and too strong for the cougar to wrench back and break them. As for the dangerous part, as docile as cows may seem, they do have a survival instinct and can and will fight back. Being stepped on by a cow or having it fall on you could injure a cougar. For a predator, an injury, even a mild one, could be life threatening. Such injuries reduce the chances of the cougar successfully capturing his next meals and lead him down the road to starvation. As a result, all predators have to take into consideration the risk of injury when selecting their prey. If that risk is high, they cross that prey off their menu. Cows are too hard and dangerous for a single cougar to kill, and because cougars normally hunt alone, they don't hunt cows.

The calves, on the other hand, are a different story. Calves are small, don't pose much danger, except for an angry mother, and are relatively easy to kill. In some parts of the Southwest, cougars have been found to be major predators of calves.[7] In other areas, for example the north-central states, including my study area, reports of cougars killing calves are rare. What are the reasons for these differences? Surely, cougars in the Southwest do not differ greatly from their northern cousins. The answer can be a complex one, but one major element is where and how the calves are raised. In northern states, spring comes late to the mountains and thus it is not worth turning cattle out into the mountain pastures until late May and early June. Ranchers want their calves to be big enough to sell by fall and so they plan on the cows giving birth to their calves in February. Because this is still late winter, this means calves are born down in the valleys, away from preferred cougar habitat in the mountains. By the time calves and their mothers reach the mountains in May and June, they are already four or five months old and can keep up with their mothers. Cows can defend their calves better against cougar attacks, and the half-grown calves are a little harder to kill. Both of these factors probably discourage cougars from trying too often to kill calves. In addition, May and June are the months when deer and elk have their young, providing an abundant natural food source for cougars, further reducing the need for them to go after calves.

In contrast, in southern mountains, snow is found only on the highest peaks, and spring arrives early. Many ranchers in this region turn their cattle out into the mountains before the calves are born. This means calves are born in cougar habitat and often before native species have their young. This is a formula for trouble: there are now a lot of wobbly-kneed calves who have trouble keeping up and being defended by their mothers. Cougars, who are evolutionarily designed to include the very young of their prey in their diet, view these calves, the only young around at the time, as easy prey. The result is significant trouble with cougar predation on calves in these regions. The basic problem with calves in the West appears to be related to the husbandry methods used in raising them. If you keep young calves and cougars separate, there should be minimal problems. Mix them together and you are asking for trouble.

The second large domestic livestock to consider is the horse and other associated equine species, for example donkeys and mules. How do horses and cougars interact? Do cougars prey on horses? The answer to this question is a qualified yes. There is no doubt that cougars attack and occasionally kill horses, with a seeming preference for colts. This appears to occur over all the West, from north to south. Most times they are not fatal attacks per se but leave the horse injured enough that it has to be euthanized. The incidents are not common but are not rare either. Because the bond between humans and their horses is often closer than with cattle, even the rare occurrence of a cougar attack will often be of a "favorite" or valued horse. Thus humans in the West often have less tolerance toward cougar attacks on horses, at any level.

But why do cougars attack horses, especially adult ones, if they are reluctant to attack similar-sized cows? The answer to that may lie in the neck again. In contrast to the short, bulky necks of cows, horses have long, elegant necks, which look a lot like those of deer and elk. I suspect that many a cougar has initiated an attack on a horse thinking it was a deer or an elk. A colt is also similar in size and probably as easy to kill as a large deer, and often cougars are successful. When attacking an adult horse, cougars are probably surprised to find the animal is bigger and faster than a deer, and the attack fails. This probably explains why many cougar attacks on adult horses result in claw marks around the neck, as a cougar fails to get up on the horse's back, or on its hind quarters, as the horse runs away faster than the cougar expected. Although sometimes adult horses injured by cougars have to be put down, most survive their injuries. It is the colts, then, that are primarily killed by cougars. Again, the level of the problem is related to how separate colts are kept from cougars.

Regarding smaller livestock, particularly sheep and goats, because these animals are well within the range of prey size for cougars, it is well established that cougars attack and kill them. Most of the problems are related to commercial sheep and goat operations where bands of a hundred or more animals are grazed in upland forested areas in the summer. In these areas they come into direct contact with cougars, and so conflicts are unavoidable. As I discuss later, the problems with predation by cougars can often be reduced by improved husbandry methods. There does seem to be an increasing problem with cougars attacking goats and sheep on "hobby" farms. Under these circumstances, the owners of goats keep them as pets or as a source of subsistence milk or meat. Overall, the majority of livestock problems regarding cougars, as well as other predators such as coyotes, then, involve sheep and goats.

Now that we know what types of livestock cougars attack and kill, we need to address just how often they do it. Do they kill hundreds, thousands, or even tens of thousands of domestic animals a year? What percentage of the total number of livestock do those numbers represent: 5 percent, 10 percent, or more? These are important questions to answer if we are to assess just how serious a problem cougars are in the West and, more importantly, if they will possibly be a problem in the prairie region. As with the risk to ourselves, this will help us determine just how much of a threat cougars might be to the livestock industry of the region.

How many do they kill? One way of looking at this issue is to examine the diet of cougars and see what percentage of that diet consists of domestic animals. The higher the percentage of livestock in their diet, the more domestic animals they are consuming and the greater the threat. There have been multitudes of studies on the diets of cougars over the years, and fortunately much of that information has been summarized by Kerry Murphy and Toni Ruth in 2010. These data come from analyzing the stomach contents of cougars killed by humans and the makeup of fecal material, "scats," of live wild cougars. The types of hair and bones left behind are unique for each prey species, so we can calculate the numbers of times the remains of each species are found. We use this as an indicator of how often each species was killed by cougars.

From these studies we know that cougars do eat a wide variety of different prey, including other cougars. Not surprisingly, deer are the most common food consumed by cougars, easily averaging over 50 percent of the diet and often as much as 90 percent. But what about livestock? When we look at what percentage of the diet of cougars consists of cattle, mainly calves, the average for the twelve western states where cougars now occur comes out to be about 2 percent. When we do the same for horses, only one study out of twenty-eight reported finding horse hair in the stomach of a cougar. Sheep and goats also only make up a small percent, about 1 percent, of the diet of cougars across all studies. Based on these diet studies, although we know cougars kill domestic livestock, they don't do it very often, or at least a majority of them don't do it.

But we know that they do indeed kill livestock, and at times a rancher can experience high losses due to cougars; reports of several sheep being killed in one night are commonly heard. Can these animals, the ones who do kill domestic livestock, inflict significant damage to the livestock industry? The way to get at that information is to look at estimates of how many domestic animals are being killed by cougars. Again, fortunately, this information is gathered, by the National Agricultural Statistics Service (NASS) of the U.S. Department of Agriculture.[8] They keep track not only of how many cattle, sheep, and goats are killed by cougars but also how many are killed by other predators and other causes.

When we look at a typical year for the decade 2000–2010, throughout the United States around 4 million head of cattle, mostly calves, die each year from various causes. Of the total that died, about 150,000 or around 3–4 percent were killed by predators. Cougars kill around 11,000 cattle, mostly calves, per year. This represents only 7 percent of the animals killed by predators and only 0.3 percent of the total 4 million that are lost. Domestic dogs kill twice as many cattle, around 25,000 per year, as cougars do. More cattle are stolen, 21,000 a year, than are killed by cougars. So we see that cougars only account for a small portion of the cattle that die or are lost yearly. When we compare it to the total number of cattle that can be found in the twelve western states, around 34 million head, the amount that cougars kill, 0.04 percent, becomes even more insignificant. So although cougars may kill over 10,000 cattle per year, this does not represent much of a threat to the cattle industry as a whole nor does it represent much of what is lost from a variety of other sources.

Similar figures appear when we look at sheep and goats. For sheep, again, ole Shep kills more than twice as many as do cougars, around 30,000 a year versus around 13,000 a year. The total number killed annually by cougars is only 2 percent of the more than 600,000 that die from all other causes and 0.4 percent of the total number of sheep in the West. For goats, the numbers are similar with about 9,000 killed by cougars, representing only 2 percent of total losses and 0.6 percent of the total number of goats in the region.

What do all these statistics and numbers tell us about the risk of cougars to livestock? What they demonstrate is that over their current range in the West, cougars represent a very minor threat to the livestock industry. There are many other, possibly avoidable, causes such as weather, theft, respiratory problems, and so forth that account for many more losses of livestock than cougar kills. Spending time and money on preventing these other losses would prove to be more beneficial than the time and effort put into control of predation by cougars. The livestock industry being concerned about cougar predation is akin to being worried about getting a cold during an outbreak of bubonic plague.

Given all the data about how cougars impact livestock where they currently occur, what can we expect with their return to the plains region? Returning cougars will most likely establish themselves in the areas less inhabited by humans, the wilder corners of the plains region. In any other places they would not be tolerated. It is in these remote areas, where livestock are most likely to roam free, where conflicts may arise. However, the overall predation rate on domestic livestock in such areas is relatively low, and I predict a similar low rate in these remote areas of the plains states. One common factor with most western cougar–livestock problems is that they are related to the husbandry practices used by the livestock owner. Often livestock owners are asking for trouble by the way they either do not know about or purposely ignore potential conflicts with cougars. Common sense would argue against letting your cattle give birth to their calves—the main product and profit of the operation—in the midst of cougar habitat. Often, simple changes in these practices can reduce or even eliminate the problem. It depends on how much a person is willing to take into consideration the elements they choose to face when deciding to graze animals in cougar country.

Even though the threat to the overall livestock industry in the Midwest will probably be minor, what about the individual rancher or farmer who, even with concerted efforts, continues to lose livestock to cougars? Sometimes these losses for individual ranchers or farmers can be relatively high. For a person losing animals to predators, statistics and advice become meaningless, and the losses are not insignificant. Overall, though, with the return of cougars to the prairie region, few ranchers should experience high losses. But how does a rancher who does have these higher losses deal with it? How should we react to these incidents? Some western states offer compensation for livestock killed by predators, especially wolves. This is done as a manner of recognizing the ranchers' contribution to maintaining a degree of "wildness" in the Wild West. Likewise, in the Midwest where wolves have returned, similar compensation plans exist. The plans are not perfect; there are complaints of abuse and of unnecessary red tape, including that the ranchers have the burden to prove that a predator killed the animal. In Sweden, where reindeer herders lose animals to wolves, bears, lynx, and wolverines, herders do not have to prove livestock loss but receive a yearly compensation based on the number of animals they have, the estimated density of predators, and the estimated number of livestock each predator species kills annually in the region. In this manner, the herders are "pre-paid" for their losses by society in support of maintaining these predators within the ecosystem. There are likely inherent problems with this or any system, but instituting such a system in the Midwest might reduce the resistance of livestock owners to the return of cougars to the region. In all cases, the compensation for losses by returning cougars to the prairie region of the United States has to be addressed in a fair and equitable manner. If we as a society judge that having a large predator within a region is desirable, the responsibility for maintaining cougars in the prairie ecosystem must not be shouldered by just one segment of that society.

A compensation system might be beneficial for the overall losses of livestock by cougars, but it still would not adequately address specific owners who might experience high losses, usually from a particular "problem" animal. In most western areas, such offending animals are often killed. Even in California, where cougars are protected by law, several hundred are killed yearly as offending animals, many for attacking livestock. In dealing with these situations, the Cougar Management Guidelines published in 2005 recommend six basic elements: (1) give a clear description of the property involved, (2) conduct an inspection of the attack site and verification of damage or loss, (3) provide owner with information on how to protect livestock, (4) issue a permit to the owner to kill the offending animal or animals, (5) allow a government agent, usually USDA Wildlife Services, to help, and (6) require the owner to report if and how many animals were removed. This same level of procedure should be established in the Midwest. This would assure the livestock owners that they have recourse for any problems that may arise. If livestock owners are either adequately compensated for occasional losses or feel that there is a logical procedure for dealing with reoccurring problems, there should be less resistance to the return of cougars to the prairie region from this important segment of midwestern society.

Although I have been talking about commercial livestock, there has been a growing problem in the West with cougars killing pets or "hobby" livestock. Many people in the West have recently moved to and beyond the edges of existing towns and cities to enjoy the rural environment. In doing so, they often move into existing cougar habitat and place their pets, mainly cats and dogs, at risk. This risk is increased if they allow their animals to run free. Many also use the opportunity of living in the country to raise hobby livestock, mainly goats and sheep (see above). As with regular pets, bringing sheep, goats, miniature horses, llamas, and the like into cougar country is inviting problems unless those animals are adequately cared for. There are also many people in the rural Midwest who have pets and hobby livestock. Because these small, deerlike animals are easy for cougars to kill, we can expect that a certain percentage of the depredation problems with cougars will involve these animals.

In California, complaints of cougars killing pets are handled in a similar manner as with livestock; a policy likewise should be implemented in the Midwest. However, many of the people living in or moving to rural areas do so because of the increased freedoms they feel such areas offer. In doing so, they should keep in mind part of the freedom of living in a wilder area involves increased risks, in this case for their pets and their livestock. If by their actions, in this case bringing in easy food for cougars, they necessitate the killing of a cougar or other predator, they are certainly contributing to the destruction of the very wildness that they desire, or to paraphrase the Joni Mitchell song, destroying nature to put "up a petting zoo." With some simple precautions, such as not allowing pets or livestock to roam free at night, such incidents can be kept to a minimum, and humans and cougars can coexist.

Although there can be reasonable plans and actions developed to tolerate cougars in the more remote regions of the midwestern states, there are many areas where their presence should not be accepted. With the return of cougars to the region, there should be by default certain suburban areas, no-cougar-lands as it were, where their presence and our tolerance to them should be zero. Luckily, most cougars seem to instinctively avoid these areas, but occasionally an animal may show up where it should not be and decisions need to be made (fig. 17). Whether offending animals are killed or translocated must be decided on a case-by-case basis. This should not be done, however, out of fear or hysteria but based on a rational plan for response to such incidences, as has been done in several western states. The key to this plan is identifying what would be appropriate actions of police or wildlife agency personnel most likely to respond initially to reports of cougars in such areas. Again, the Cougar Management Guidelines provide information on the types of things to consider in setting up a first responder protocol for handling cougars in these situations. The details of such protocols are too extensive to outline here, and I refer you to this excellent publication. The overall idea is to first assess the risk level, that is, is it just a sighting or a more aggressive encounter, and then figure out the appropriate response. One important element of an appropriate first response is for the people involved to have some understanding of cougar behavior and basic handling techniques. There are several sources of information and training available to obtain this understanding. The California Department of Fish and Game has developed a First Responder DVD in which they outline the procedures they use in such situations.[9] The Cougar Rewilding Foundation offers hands-on workshops for first responders led by Dr. Jay Tischendorf, a noted wildlife veterinarian.[10] The workshop includes learning about the basic biology of cougars, what precautions are needed when dealing with them, how to tranquilize them, and even how to deal with the public in such situations. This and other workshops will likely grow in number as personnel of state agencies in the Midwest require more training to deal with the increasing occurrence of cougars in the region. Such training will reduce the likelihood that when cougars show up where they are not supposed to be, actions will not be the results of hastily drawn-up plans based on fear and ignorance. If the animal is to be killed, it should be done only after careful consideration of all options. We need to remember that cougars are not inherently aggressive toward humans; they do not come hundreds of miles seeking us out. This is especially true of an animal that has wandered into an area where it should not be. In such situations, the cat is often more scared than any humans and the last thing on his mind is to attack. In fact, there has never been a reported incident of cougars in these situations attacking. Unfortunately, many agencies and news personnel—for sensational purposes—often portray these cats as being highly dangerous, but the reality is far from the hype, and these situations should be handled accordingly.

In summary, regarding both the potential interactions of cougars with humans and their livestock in the Midwest, past and present experiences with existing cougar populations indicate that there is no need for hysterical alarm. Cougars are not that dangerous to us and even less so when compared to other risks we face daily. Cougars will also pose only a minor threat to the livestock industry of the region, and with well-conceived and well-conducted compensation/response programs, special cases of problem animals can be handled. Later in this chapter we look at all of the information I have presented and discuss whether it is being or will be taken into consideration. But first, we need to consider the final area of potential conflict between cougars and humans regarding the wild livestock in the region: the deer.

Deer, mainly white-tailed deer, in the prairie region are also a vital component of the original prairie ecosystem. As mentioned earlier, an ecosystem functions by the flow of energy through it, and deer, as a major herbivore, participate in converting plant energy into animal tissue, originally to support top predators, including humans. In the original prairie ecosystem, deer were not as plentiful as bison but were in significant enough numbers to support the populations of river lions. As with all the other large wild mammals, deer too were extirpated from much of the plains region when it was settled. However, the recovery of deer populations, both white-tailed in the East and mule deer in the West, is one of the wildlife success stories. In many states where there

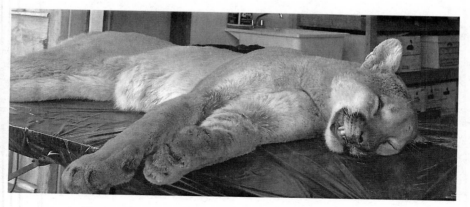

FIGURE 17. Young male cougar shot and killed in central Missouri, exemplifying the fate of most cougars who show up in urban or suburban areas of the Midwest. (Photo courtesy of Don Shrubshell/*Columbia Daily Tribune.*)

were virtually no deer only eighty years ago, they now number in the hundreds of thousands. Current deer populations are now so high that many people are concerned that we might have too many deer. They have become menaces on the highway, causing billions of dollars of damage and hundreds of human deaths. As with any organism, including humans, high densities tend to facilitate the transmission of diseases. In this case, it is not surprising that Chronic Wasting Disease (CWD) in deer has reached epidemic levels in many areas and appears to be related to deer densities. The increased incidence of Lyme disease in humans, transmitted by the deer tick, also seems to coincidently be related to high deer populations.[11] High deer densities, especially in the absence of its top predator, can have major impacts on forest regeneration and biodiversity.[12]

In light of all these problems associated with what many feel is excess deer numbers, one would think that the return of the cougar would be welcome. With its return, we might hope for lower deer numbers or at least for deer being less domestic within a reestablished landscape of fear. The recovery of deer populations owes much of its success to the fact that many people *wanted* deer to return, mainly to hunt them. Because most deer hunters want to be successful, few within the deer-hunting community believe there can be too many deer. The return of cougars to the Midwest does not appear as a solution to what many consider a nonexistent problem but a threat to their recreational goals. They consider any decrease in deer numbers to be undesirable and often quickly blame state game agencies when such declines, no matter how small, occur. Yearly deer harvest numbers are compared like team win-lose records to see how Team Deer Hunter did. Any decrease in harvest is seen as a losing season and suspicion is cast upon the coach (game agencies). In the Midwest, deer hunters cannot currently blame cougars for declines in deer; in the West, however, any change in deer numbers is quickly blamed on cougars, and now wolves, taking the pressure off game agencies. Thus, the hunting community in general does not look favorably on the return of cougars to the Midwest.

On the other hand, are hunters justified in their belief that cougars would reduce deer numbers sufficiently to affect their sport? And for that matter, what evidence do we have that cougars returning to the Midwest could effectively control deer numbers and provide all the hoped-for benefits? These are two contradictory views of deer numbers, and they both have in common the concept, the fear or hope, that cougar predation can have a significant impact on deer populations.

Since deer numbers, either too many or not enough, seem to be a concern in the plains region and could affect the social acceptability of cougars returning, we need to try to answer a basic question: Can cougars significantly impact deer numbers? Unfortunately, once asked, it becomes a difficult question to answer. Needless to say, because of the concern of hunters, who ultimately pay for most of the wildlife research conducted in the United States, this question has been studied a lot. There have been literally hundreds of studies, published and unpublished, directed toward this question, and not just concerning cougars. Because predators in general tend to eat the same species we like to hunt, the concern that maybe those predators are eating too many of our game is a dominant one in wildlife management. We have *theoretical* studies, whose complex mathematical formulas attempt to incorporate obvious and not so obvious details of predator–prey interactions. We have *lab* studies where insects and other small invertebrates (and even larger animals) are thrown together to try and tease out the impacts of predation. We have *field* studies that attempt to manipulate mostly predator numbers to determine if such manipulations result in more or less prey. We have studies that combine all three approaches, but after all these studies, what do we know? Let's return our discussion to deer and its predators, primarily wolves, coyotes, and of course cougars. Again, what can we say for sure? As summarized by G. E. Connolly in 1978 in his review of coyote predation on deer, by selecting the right literature, one could find support for any or all views of the role of predation. So in 1978, the answer to the question was yes or no or maybe or we don't know.

Since 1978 there have been a large number of new studies with more sophisticated techniques, larger samples, all the stuff that good science is made of. Have all these new studies refined the answer? In 2001 Warren Ballard and others repeated Connolly's exercises and reviewed more than thirty newer studies of the impacts of coyotes and cougars on white-tailed and mule deer.[13] They incorporated various theoretical ideas regarding carrying capacity, compensatory and additive predation, limiting versus regulating factors, all the latest. What did they find? Do predators, cougars included, have an impact on deer numbers? Again, the basic consensus, depending on the study chosen, was that one could come to whatever conclusion one wanted. Only on certain small (fewer than 1,000 square kilometers) areas did predator removal result in increased deer numbers, meaning that predators were assumed to be keeping deer numbers down before removal, maybe.

All these studies demonstrate that there is no solid evidence that predators in general and cougars specifically can have significant impacts on deer populations. Articles since 2001 continue to reinforce that opinion. Accounts of teaming herds of ungulates and their predators in precolonial America reinforce that opinion. The vast herds of wildebeest, zebras, and others roaming the plains of Africa with lions, wild dogs, leopards, and cheetahs reinforce that opinion. Isle Royale National Park in Lake Superior with one of the highest densities of moose and wolves in North America reinforces that opinion. And this is an opinion that makes logical evolutionary sense. Any predator that would have a detrimental impact on its prey would have quickly died out with its prey eons ago. Only the slightly inefficient predators can survive in the predator–prey game. To be *super* efficient is evolutionary death. What results, if not disturbed by humans, is a system where prey drive the predator, not the other way around. As prey numbers change for other reasons, predators track them, increasing during good times, starving when prey become few and far between during bad times. That is how ecosystems work; energy flows upward.

What, then, causes changes in prey numbers? If we look at most of the early work in deer population biology before the 1980s, many of the investigations were looking at the impact of weather on deer numbers. At that time, except for coyotes, there were probably few other large predators to blame for changes in deer numbers. Yet deer numbers still fluctuated, much to the distress of hunters and game managers. At that time hunters blamed these declines on poor management by game agencies, pointing to allowing them to hunt female deer, not providing winter food, and so on. Not surprisingly, investigators did find relationships between unusual weather, for example deep snow in the winter or drought in the summer, and deer numbers. Often there were extreme fluctuations in deer numbers in response to weather because during good times, deer numbers increase well beyond what an area could support in bad times. So before the 1980s, many wildlife biologists concluded that weather was a major driving force in deer numbers, not so much in controlling or regulating them, but in limiting them based on what an area could support under the worst conditions. Starvation, especially in the young, was recognized as a major mortality force that periodically reduced deer populations.

Maybe because we can't do anything about the weather or maybe because it is easier to believe that since predators kill prey they should have an impact, studies began to emphasize that it was not so much the weather but, in reality, predation. They talked less and less about the weather and more about predators. In fact, most modern predator–prey studies don't even consider the weather; they assume it is predation and set out to prove it. But, as demonstrated by others, after thirty years of trying, that connection has yet to be made. In science we are supposed to operate on this basis: once you show something to be incorrect, you discard that idea and move on. It is only when you find support for the idea that you continue to test it, trying anew to prove it erroneous. It appears that regarding predator effects on deer, even though it has been refuted many times, we continue to test it in hopes of proving it right.

If it is not predators, then what is it? What causes deer numbers to decline, often precipitously, from one year to the next? Again, we have to return to the weather. As I demonstrated with cougars and mule deer in Idaho, when we considered predation by cougars on a severely declined deer herd, deer numbers would be predicted to rebound to pre-decline levels within five to six years, under average winter precipitation conditions and high numbers of cougars.[14] In reality, though, deer numbers stayed low and bounced up and down because winter precipitation bounced up and down. By incorporating this varying precipitation, in this case snow depth, predicted deer numbers immediately replicated the pattern seen in the field data. Cursory observations of other areas—desert areas in the south, snowy mountainous ones in the north—again show a general pattern of increasing deer numbers under good weather conditions and declining ones when conditions are bad.

What is the mechanism of this impact of weather on deer numbers? Data from various studies, in particular from those in Colorado, demonstrate the strong influence of snow depth on fawn and female deer survival.[15] As basic population biology tells us, the young control the growth of a population. Increase their survival and the population grows; decrease their survival below replacement and the population declines. Though predators such as cougars can kill fawns, they have to find them to do it, and studies show that they are not very good at it. Fawns cannot hide from starvation; it silently seeks them out wherever their beds are concealed. This silent killer may not be as dramatic as a cougar or coyote killing and eating a fawn but is more deadly on a population level. Deer populations, then, as they have for millennia, will continue to fluctuate wildly as dictated by weather.

How does all this pertain to the Midwest? Though the arguments will continue to rage regarding the impact of cougars on deer populations, the evidence indicates that cougars returning to the plains region will have minimal lethal impacts on deer populations. I base this conclusion not only on the lack of evidence that they have significant lethal impacts on deer but because of several other factors unique to the plains region regarding deer. First of all, most midwestern states have hundreds of thousands of deer across their states. Deer hunters per state kill as many as 100,000 deer, males and females, year after year. Given my earlier estimate of a maximum of 1,000 cougars that would likely return to the plains region, this number is not enough to make a dent in the vast deer herd living across this region. For example, if we assume one adult cougar kills on average a high estimate of forty deer per year, this would mean all 1,000 cougars in the Midwest would kill only 40,000 deer a year. This is less than half of the number that deer hunters in one state kill per year. Spread over the millions of deer living in the Midwest, this lethal impact would be minimal.

A second reason that cougars will not have a detrimental impact on deer populations in the plains region is that a majority of the deer found in this region live in fairly intensively farmed landscapes. These deer flourish by finding cover in the small woodlots and remaining river bottoms while feeding on the abundant corn and soybeans in the upland areas, similar to the ways their ancestors did on the native prairie species. It is in these vast areas where the tolerance for the presence of cougars will be the lowest. Individuals, let alone populations, will not be permitted to establish themselves in these areas. As in precolonial times, there will be a vast reservoir of deer in areas not frequented by cougars. This protected population will continue to reproduce and replace any losses from hunting, either by humans or by cougars.

That said, does this not negate my earlier argument that we should allow cougars to return to the plains region to reestablish the ecological balance, to relieve the pressure of too abundant and too tame deer on a suffering ecosystem? In these huge, intensively cultivated regions, there is little to save and there is no silver bullet solution such as a top predator that can protect what little remains. The small fragments of native vegetation, upland prairie, or lowland river forests are, in most cases, just that, fragments, relics of the once mighty prairie ecosystem that covered that region. They are like pieces in a museum or exhibits in a zoo; they can provide the viewer with a glimpse of a past culture or a wild animal, but they are not and cannot be functional pieces of the ecosystem. These areas might be refuges for some of the original prairie components, such as the prairie bush clover (*Lespedeza leptostachys*) or the purple fringed orchid (*Platanthera psycordes*). With intensive efforts such species can be conserved, but the ecosystem itself is gone.

Cougars can make a difference in the areas where there is still room for them to live, those wilder places that still exist. It is in these areas, as with the precolonial river breaks, that cougars can reestablish an ecological order, can help make these areas more than just large relics. They can make them living examples of the lost paradise once called the Great Plains. They do this not by their lethal impact on deer, which as we saw is minimal, but in the same way their ancestors did: through fear. By reestablishing the landscape of fear for deer in these few but significant areas, such as the sand hills of Nebraska or the

Badlands of North and South Dakota, these relatively few cougars will wield a far greater control over the deer than just by killing them. They will keep deer in their place, make them wary to go where they could get killed, make them look over their shoulders, make them wild again. And in this wildness, there will again be refuges for plants, and subsequently animals, typical of the prairie region whose last hope of survival are these wilder areas currently threatened from overuse by deer. With the return of cougars to these areas, there will be a rebirth of diversity unparalleled in over one hundred years. Although cougars cannot rescue the many small islands of prairie habitat, they can help the salvageable areas and return them to their functional states.

In summary, the overall answer to whether the return of cougars to the plains will have significant impacts on deer numbers: no and yes. There will be too few cougars returning to too few areas in the Midwest to have any consequence on the overall deer populations within those states. Hunters need not be concerned; there will be plenty of deer to kill, depending on the weather. But, in the areas of the prairie where sufficient prairie habitat still exists, the return of cougars to these regions will provide the relief needed from too many and too tame deer, returning them to a more functional prairie ecosystem. These areas represent a very small percentage of the original vast prairie ecosystem—they are what is left. We are indebted to future generations to save these lands, and cougars can help us in these endeavors.

Having established that cougars are not that big of a threat to ourselves or our livestock, domestic or wild, how should people of the Midwest react to the return of cougars to the heartland? If taken rationally and calmly, I propose that every state can develop a reasonable management plan of action regarding both the movement of cougars through their state and the possible establishment of a resident population within the state. The first step in the management of any wildlife species is to first know where that species is most likely to be found. As outlined in chapter 4, each state has a different potential for cougars establishing breeding populations within their boundaries. South Dakota has a higher potential to have cougars out on the prairie because more areas of native vegetation exist in the state and because the Black Hills already has a resident population. A state such as Iowa, however, has few unsettled areas and so its overall potential is much lower. Even within a state, the possibility of resident animals changes as one moves to more cultivated regions. Thus, each state should conduct its own analysis similar to what was done in North Dakota. That report is an excellent example of assessing potential habitat for cougars within the state. I made my assessments based on what I know about cougars and a bird's-eye view of the landscape. This may be a good place to start, but people with more familiarity with their state's landscape should do their own evaluation, as was done in North Dakota. This would produce a cougar probability map for each part of the state. Such a map can be helpful in deciding where resources are most and least needed in dealing with cougars. State personnel can first of all identify where people would most likely not see cougars. These regions then would not need as much emphasis as other regions. In areas where cougars might pass through or establish themselves, state agencies can begin to inform people of these possibilities. They could provide seminars and workshops designed to increase the knowledge of cougar biology among the general population. They could inform people of the potential of having cougars in the area and how they should react and deal with any situations that arise. Special emphasis could be placed on providing first-responder training to local law enforcement personnel so that any situation would be handled in a professional manner. Using the Cougar Management Guidelines as a model, each state can and should have in place a set of procedures designed to handle most possible incidents.

There is no reason why people in the Midwest, given the same information on cougars, should respond any differently than people in the more western states do. Information is the greatest deterrent to hysteria and panic, and the more information people have regarding cougars in the areas where they may show up, the more likely the local people will react in a calm and reasonable manner. Even though the outcome of some of these incidents will result in the killing of a cougar, at least it will be the result of a clear and logical evaluation and not done in a hastily fired rain of bullets.

In areas where cougars are likely to establish populations, plans should be made regarding any conflicts with livestock. Ideally, a response/compensation program should be in place before cougars arrive. In this manner, again, any situation can be handled in a rational way. These programs should be built around the concept of adaptive management, which is designed to allow changes in management plans as new data and information are collected.[16]

In the development and planning of all these responses, the cooperation of all the stakeholders involved is needed. Citizens need to be informed of the real dangers involved and assured that tolerance of cougars in the region does not involve unwarranted risk to themselves. Citizens concerned about the welfare of wildlife, especially that of the cougar, need to be reassured that their desire to see the return of a charismatic species will be considered. Ranchers and farmers need to be assured that, though the overall risks are small, there is a reasonable plan in place to safeguard their livelihood. The hunting community must be willing to accept the role of cougars in the ecosystem but also must be assured that deer populations will not decline with the cougar's return. Finally, politicians need to be willing to work with all constituents and not use the return of cougars into the state as a tool to generate fear and suspicion for their political gain. If all members of a state work together in good faith and with anticipation, there is the possibility that the return of cougars to their state will not present major problems or concerns. These may seem like lofty goals regarding how society should react to the return of cougars, but we need to set our sights high. We need to rise above the unfounded fears, the uncontrolled reactions, if we are to hope that cougars can be socially and politically integrated into the Midwest. The ultimate question is: Will we rise to the occasion?

That is how, in my opinion, people of the Midwest should react to cougars returning to their region. As we all know, there is a big difference between how one thinks people should react and how they actually do react. So the next question is: How are the people of the Midwest going to react to the arrival or pending arrival of cougars to their region? This is actually a multistep question because it seems to depend on whether cougars have actually returned or not. From what I can see based on headlines and comments across the Midwest, there are several stages that people go through. I will use the Black Hills of South Dakota as an example of what I am seeing and expect to see across the plains. As mentioned, cougars were extirpated from the Black Hills in the late 1800s and became established again in the late 1900s. During that time, there were periodic unconfirmed sightings of cougars, and most people did not believe them. Not having to face the actuality of cougars being there, in general, peoples' attitudes were basically benign.

Once cougars began to show up in the Black Hills, first with verified sightings and then with documentation of a breeding population, there appeared to be an overall enthusiasm that such an event could happen. The attitudes of people aligned with nature groups seemed to prevail with a general consensus that this was a positive thing. To its credit, the South Dakota Game, Fish & Parks (SDGFP) Department responded in a reasonable manner. They developed a departmental guiding philosophy regarding the return of cougars. In that philosophy they stressed that mountain lions contribute to the quality of life in South Dakota and that they play an important role in the ecosystem. They further stated that mountain lions should be managed in accordance with biologically sound principles, that education and involvement strategies with regards to safety should be implemented, that accurate and timely information should be provided the public, and that their agency and the public have the responsibility to learn to live with mountain lions in order to maintain viable populations while dealing with problems that mountain lions might cause. They ended by stating that the future of cougars in South Dakota depended on a public that appreciates, understands, and supports mountain lions.[17] Starting in 2002, the SDGFP Division of Wildlife conducted a series of public opinion surveys and held various public meetings.[18] In one survey of more than four thousand people of the general population of the Black Hills, more than 56 percent favored either current levels or increases in cougar numbers in the region.[19] In this same survey, when asked what actions should be taken if a cougar frequented an area where they lived but did not cause problems, over 90 percent of the respondents said that the animal should either be captured and removed (30.6 percent), chased away (10.6 percent), the public should be educated on how to live safely (38.5 percent), or take no action (11.7 percent). Only 7 percent said the lion should be killed.[20] The public also agreed (71.1 percent) that a season for hunting mountain lions was acceptable, if the state could verify that the population was healthy and could sustain a prescribed level of harvest.

The South Dakota Game, Fish & Parks Department also initiated a study of cougar ecology and behavior in the Black Hills where they attempted to address basic biological questions regarding how many cougars there were, what they were eating, and how they were using the Black Hills area. Preliminary data of these studies indicated that cougars were indeed feeding mainly on deer but that there was no evidence that they were affecting the deer population. They also estimated that the cougars were probably around carrying capacity at about one hundred thirty or so adult animals.

Initially the department seemed to have the backing of a majority of the public for a sound and sensible management of cougars in the Black Hills. They also had an increasing data set that documented that the population was healthy and not causing major problems for human safety (no reported incidents involving cougars and humans), livestock (no domestic livestock have shown up in their diets), or deer. Because the population appeared to be healthy and could sustain a certain level of hunting, in 2005 a hunting season was initiated with a total quota of cougars set at twenty-five and a subquota of five adult females. That first year thirteen cougars were killed. If the population of adult animals was around 130 at that time, this would represent around a 10 percent mortality rate, which by all standards should be supportable.[21] The following year the quota stayed the same but it increased to thirty-five the next two years, and in

2010 it was set at fifty, or about 40 percent of the estimated adult population. A hunting mortality rate of 40 percent plus additional natural mortality is commonly recognized as not being sustainable without severely decreasing the total population. In 2011, the quota was raised to seventy, with the Game, Fish & Parks Department admitting that they did indeed want to reduce the cougar population. Thus it appears that in only six years, South Dakota went from what appeared to be a logical, sensible approach to managing its new cougar population to one that appears to be set to decimate that population and possibly eliminate it from the region.

One has to ask what happened over those six years. How and why did the game agency move from its original philosophy and the expressed will of the citizens to maintain a healthy cougar population to one that is not based on any sound modern management strategy for a wildlife population? Proposing a hunting mortality rate of 40–50 percent for a deer population consisting of tens of thousands of animals would never even be considered, but yet it is being urged for a population of only, at a maximum, 130 adult cougars. The Black Hills has a population of around 160 bighorn sheep, and the game agency would never consider letting 50 percent or over eighty of these animals, males and females, be killed by hunters. Yet it seems acceptable for the cougar population. The population of Florida panthers, a subspecies of the cougars in the Black Hills, is only around 120 animals and is considered highly endangered. Florida game management personnel would not consider a hunting season, let alone one where 50 percent were killed. On the other hand, there is concern when perhaps ten to fifteen animals get killed on the highways. Where is the same concern for the cougar population in the Black Hills, which by all conservation standards is endangered in South Dakota?

How has this happened? Was there a major change in personnel in the South Dakota Game, Fish & Parks Department—employees who favored sensible biologically supported management were replaced by others who thought otherwise? No, the same people who were in the agency earlier are still there. What then has changed? To understand this shift in policy and how it came about, we need to look at some later attitude surveys the department conducted and at the basic structure and control of wildlife management within the state.

In 2005 and again in 2010, the Game, Fish & Parks Department conducted public meetings specifically dealing with cougars and the possible question of having a hunting season on them. These meetings were held in a variety of towns and cities, and in each year around 350 persons responded to the same questionnaire given to the general public (4,381 respondents). Not surprisingly, the response of these more select groups, usually representing the hunting community, overwhelmingly (over 86 percent) favored a hunting season for cougars. They also favored reductions in cougars in the Black Hills by more than 66 percent. An additional survey, specifically of deer hunters in 2002 in the Black Hills, not surprisingly strongly favored a hunting season for cougars by more than 87 percent.

What these figures tell us is that the general public favored a more cautious and reasonable approach to the management of cougars. But groups admittedly biased against predators in general and cougars in particular would prefer that a species they see as a competitor be significantly reduced. Given these results, why would the Game, Fish & Parks Department put more weight in what these selective groups would want than the general public? The answer to that demonstrates the weakness of managing wildlife species, especially predator species, in the United States. That weakness stems from four factors: (1) the goal and mission of state wildlife resource agencies, (2) the political controls placed on these agencies, (3) hunters themselves, and (4) the hunting industry they have become a part of.

Game agencies in most states, including South Dakota, are funded almost entirely by license revenues from hunters and fishermen. Many agencies have changed their names from "game" to "wildlife" agencies, and have now been put in charge of all wildlife species. However, this is a cosmetic name change at best. Because they are paid by a clientele primarily interested only in the species they can hunt, the agencies' main concern is still the traditional game (i.e., hunted) species. This is reflected in the overall mission of most game agencies. This mission was concisely summarized by a former director of the Idaho Department of Fish and Game at one of these "public" meetings back in the 1980s. The director summed up all the efforts of a game agency with this simple statement: "The goal of this department is to put more fish in the creel and more game in the bag."

Others may not be as open and frank as this fellow, but that goal is implicit in the funding and operations of all game agencies. They answer directly to the hunters/fishermen, and their job is to provide, if not ensure, the opportunity of success to the sportsman. With that one statement, it became clear to me that hunting in America is no longer an outdoor adventure, no longer a test of predatory skill, no longer a contest of cunning between hunter and hunted. Hunting in modern American is akin to shopping. We go to the store expecting the shelves to be full of what we want and at the affordable price of time and energy. We expect to leave with something. Game personnel in the state agencies are the managers of those stores, and they are expected to keep the shelves stocked, to make sure the customer is satisfied. In fact, a lot of time, energy, and money are spent surveying hunters after the hunting/shopping season to gauge the level of customer satisfaction. The goal is to make sure there is "more game in the bag."

Given this type of dedicated funding and vested interest by the people supplying the funding, it is not surprising that SDGFP puts more weight in what a smaller percentage of people with vested interests have to say than the general public of the state. But that is just one part of the problem.

Even given the pressure from hunters to reduce their competitors, again to their credit, personnel of the department still stressed sound management of the population and caution concerning levels of excess hunting. However, herein enters the second element of what is wrong with wildlife management in the United States: the state game commission. In most states, besides the wildlife agency entrusted with the duty of managing all wildlife resources of the state, a second layer of administration or control is a board of commissioners that is supposed to represent the public's interest in these resources. It was originally conceived as one of those checks and balances, an attempt by the populace, including the hunting community, to have some say in how and what game agencies did. These commissioner positions are appointed by the governor of each state and thus are susceptible to abuse. What usually happens, especially in the more western states, is that many, and often controlling, positions are filled by people with other vested interests, such as large landowners and people from the livestock industry. These persons, who have substantial political clout in rural states such as South Dakota, are appointed by governors they supported and who use the position to wield tremendous pressure on the game agencies they oversee.

People in the livestock industry in general do not like predators who kill their animals, nor for that matter, ungulates who compete for their animals' forage. Thus, as often is the case, game commissions do not normally reflect the will of the general public or of the hunting community. In the case of South Dakota, the impact of this biased commission system was clearly seen in the establishment of the kill limits for the 2010 and 2011 cougar seasons. Game, Fish & Parks personnel, under pressure from hunting groups and the commission, established what they felt was as high of a quota they could in 2010 (forty animals). The commission publicly expressed a desire to raise that number. They held a series of public hearings where, in general, the people expressed concern about the killing of forty animals, let alone more. Though the commissioners patiently sat through these hearings, it was indeed a waste of time. These commissioners had already made up their minds to raise the quota by five and to add five more animals to be killed within the Custer State Park, where hunting had been previously prohibited. The next year they forwent the charade of a public hearing and unilaterally decided behind closed doors to increase the quota to seventy.

I argue that it is for actions like this that the old-style game commission system should be abolished. When America was more rural and a larger number of us hunted for sustenance, these commissions might have represented such interests. Today, however, these special interest groups do not represent the public in general and often do not represent the hunting community.

Speaking of the hunting community, this brings us to the third problem with how wildlife is managed in the United States—the hunters themselves. I come from a rural hunting tradition, and on the farm in Wisconsin we looked forward to opening day of hunting. It was a chance to get out into the field and woods, to work with the dogs, to enjoy the hunting heritage all humans have. Over the years, I have had the opportunity to kill really big deer and elk but also to walk away from the field empty handed. Back then hunting was not shopping; it was hunting. Modern hunters have been swept up into the frenzy of hunting, almost the competition of it. They are not taught it is the thrill of the chase that is the enjoyment of hunting; instead it has become the thrill of the kill, of the success. No kill means no success, no enjoyment. Hunting is no longer learned by following along with an adult, a father or uncle, and learning from example. Today it is not learned, it is taught. I heard an example of this "hunter education" on a local public radio station where young hunters were taught how to hunt pheasants. The first step in this process is to take pen-raised birds out to the training field the morning of the "lesson." These birds, which have never been in the wild, are positioned in various locations, and to ensure that they stay there, are spun around upside down to make them dizzy. They are then placed on the ground, their heads tucked under their wings, and left to await their fate. The young students are then taken out to where these dazed birds are, and one student is chosen to approach the bird, flush it, and try to shoot it. If he or she is successful, there are cheers all around. If they miss, they go in the back of the line to wait a second turn. This all sounds more like hunting for Easter eggs than hunting.

Maybe it is just that I am getting old and grumpy, but I find this type of "hunter training" disgusting. It has reduced hunting to the level of shopping or of a competitive sport. If you make a goal you are a hero; if you miss you are a failure and go to the back of the line. There is no fair chase, no enjoyment of the outdoor experience, no "sport" involved in shooting a dizzy bird that has no idea where to flee. If these types of hunter-training courses predominate across the country, it is no wonder we are raising generations of "hunters" who do not understand the concept of what hunting is or what its ecological role is. Because hunting *is* predation, the foundation of its training should be the understanding and appreciation of the fundamentals of predator–prey relationships, regardless of who or what the predator is. It is not surprising, then, that many hunters view other predators as competition, even unfair competition, for the game they need for success. It becomes a competitive outdoor sport between "us" and "them." As with many outdoor sports, we have stripped all that is good and enjoyable about hunting and made it a contest. It is a sad commentary that modern hunting is not the hunting of my forefathers or of my youth. I miss it.

What has driven this conversion of hunting into some type of competitive sport? Here the fourth element, the hunting industry, comes in. As with so many other sports, not only do the competitors strive to win but there is a whole industry focused on helping them to achieve that win. When I was young, a hunting store was small; it contained the basic guns and supplies you needed to go afield. There may have been three or four models of shotguns and rifles to choose from, three or four brands of ammunition. Maybe they sold hunting clothes and other basic hunting paraphernalia. The stores had a warmth about them, they were locally owned, and you probably knew the owner. However, somewhere along the line, the hunting manufacturers discovered marketing. They discovered that if you advertise that a hunter really needs the best gun, the best ammunition, the best clothes, they will come and buy. There are no longer a few types of bullets that seem to work well overall. Now you need specially charged, balanced bullets for each species you hunt. Ballistics of bullets has become a common consideration and concern, just like the stats of a football player. There are guns with scopes, with laser sights to reduce the chance that you will miss and thus fail. There are range finders to accurately judge distances, so you don't miss and fail. The types and array of shotguns and rifles has become a wonder to behold. No longer are there just a few favored types, a manual pump shotgun, a .30-30 or .30-06 rifle. Now there are all kinds of calibers for all kinds of hunting from big game to "varmints." You can even find automatic assault rifles, as if anyone would want to launch a full assault on an unsuspecting deer. The number and types of guns and ammunition we have available for the public, both hunting and nonhunting, easily exceeds the military supply of many countries. We have been convinced that we need to be armed to the teeth, to be successful, in order to win at this hunting sport.

The modern hunting industry has taken even the more primitive forms of hunting, bow and arrow, and muzzle-loading rifles, to new technological heights. No longer is the old recurve bow good enough. You need to have the latest compound, double action, whiz-bang bow with laser sights and special alloy computer-balanced arrows tipped with triple hooked-back razor Teflon-coated blades. Or, better yet, a modern crossbow that looks more like a gun than a bow. Remember, you can't afford to miss!

Hunting stores have become department stores, covering large spaces with the latest, must-have hunting toys. They sell trail cameras that you can install over a bait to see where the big ones are, at least before the season. They sell liquids and sprays to attract the big buck, thinking he will have an opportunity for romance just to get blown away. They sell sprays to cover your body odor. They sell camouflage clothes so you blend into the forest, making you snipers rather than hunters. They sell everything an ingenious inventor can think of to improve the chances of success. The modern hunting industry has found the modern hunter, driven by the need to succeed, to be gullible and with very deep pockets.

How have they done this? How have they worked hunters into a buying frenzy? Good old marketing! In the past we had magazines such as *Field and Stream* or *Outdoor Life* to provide us stories of the hunts where hunters were skillful, or lucky, enough to bag the big one. Today, although those magazines still exist, the flavor and the multitude of new magazines have changed. First, there are big flashy ads screaming out to us to buy the latest, the best equipment. Next there are articles on the latest and best equipment, and why we really *really* need it. There are articles on where to go to find the biggest animals with the least amount of investment in time, including the latest state statistics of hunter success, and game population levels. There are ads from guiding firms guaranteeing your success. Because of all this, hunting is no longer a sport, it is an industry, and the success of that industry depends on the success of the hunters. You cannot sell the latest super-powerful, super-accurate rifle if the buyer never gets a chance to kill something with it. For the sake of the industry, we need all the available game animals and cannot afford to have other predators, who don't buy guns, eat into the profits of this industry.

These four factors—a limited clientele of the wildlife agencies, a biased commission system, competitive hunters, and industrial hunting—result in efforts to maximize the number of game animals available, and one way to do this is by reducing or eliminating the competition, in this case the cougar. It is this flawed system of wildlife management in America that not only contributes to the sacrifice of predators for special interests and corporate profits but also excludes a majority of the citizens from having a meaningful say in how wildlife resources are managed. There are only an estimated 12.5 million hunters in the United States, or only 5 percent of citizens of hunting age. Yet this 5 percent not only make the decisions on how many wildlife species, especially the predators, are managed but feel it is their exclusive right to do so.

Thus it becomes clear why the South Dakota Game, Fish & Parks Commission sides with hunters, who are their employers, and not the general public. Regardless of what the general public may say, the hunting community can argue that they pay the bills for wildlife: if you don't pay, you don't say. Concerns by nonhunting groups such as the Black Hills Mountain Lion Society or citizens in general carry little weight.[22]

This is the way it is these days. But should it be so? Should only hunters have a say in how wildlife is managed? Are they really the only ones who pay for wildlife? More fundamentally, the question is: Who owns the wildlife of the United States and should they not have a say, whether they pay or not? These are all questions that would require an additional book in itself, and there have been several written on these topics.[23] In a briefer response here, the easiest question to answer is that of ownership. It is well accepted and written in law that all the citizens of the United States own the wildlife. Ownership of nonmigratory species officially belongs to the citizens of the individual states. This right was specifically guaranteed to avoid the European system, where only landowners had that right, and to ensure all citizens had a say in the use of these resources. To argue that only the hunters who pay via their licenses have a say in the way wildlife is managed is akin to saying that just because someone pays for the food he buys, he is the only one who has a right to say how the store is run. We would all concede that it is the owner of the store that has the say, not the clients. Hunters are clients buying the privilege to hunt from the "store" owned by all of us. Since we all own the wildlife and the public land that feeds and shelters that wildlife, should we not all have a say in how this is managed? So by law and by common sense, we all own the wildlife, and we all have a say in how it should be managed.

I would argue that we all also contribute to the support of the wildlife. We may not pay directly to game agencies via the purchase of a hunting license or excise taxes on hunting equipment, but sufficient general tax funds from both state and federal budgets support and pay for wildlife resources in countless ways. We not only pay for wildlife but also support hunting of that wildlife by taxes used to maintain the roads that game agency personnel and hunters drive their trucks on. We pay via subsides to the oil industry for the gas they use. We pay in support of public education that trains the wildlife managers. The list of this hidden support goes on and on.

In much of the West, including the Black Hills of South Dakota, our taxes to the U.S. Forest Service and Bureau of Land Management provide the public land, the nurturing ecosystem that grows the wildlife, including the cougars, that hunters pursue. We also pay to support the U.S. Fish and Wildlife Service, which manages the wildlife. We pay the U.S. Department of Agriculture's Animal and Plant Health Inspection Service's Wildlife Damage Management to kill wildlife. In all, we pay a minimum of $10 billion annually either directly or indirectly for the support of wildlife in general and game species in particular. Although it would be nice if nonhunters who enjoy the outdoors would also pay extra via an excise tax on all sporting goods, we already pay several times more than the hunters do with their license fees or excise tax on their hunting equipment. So not only do we all own the wildlife, but we also all pay for it.

What does this mean for the cougars in South Dakota, or the rest of the Midwest? It means that the game commissions of these states have not only a moral but also a monetary responsibility to listen to the general public. It is not just limited vested interest groups that pay the bills; it is the majority of the citizens of the state. If game commissions cannot be the voice of all citizens, then they should be abolished and replaced with some mechanism that will represent all citizens of the state. Additionally, I argue that state game agencies should be divested of their role in managing all the other "nongame" wildlife, including predators. They only manage (kill) predators to protect favored game species. State game agencies have amply demonstrated that they are not paid to and thus have no interest in managing nongame species. So perhaps it would be better to let them manage their ducks and deer but create a separate agency for the rest of the wildlife. Not the nongame programs we currently have but full-fledged state departments that will be directly responsible to the people who pay the bills for all wildlife: the public. At the very least, the management and fate of predators in general and cougars in particular should not be in the hands of a small (less than 10 percent) number of citizens whose main interest is to have them eliminated.

In summary, the general pattern exemplified by the South Dakota example is one where consistently the general public favors a rational and balanced approach to the management of cougars in the state. The state game agency initially acts on these wishes and begins to develop a management strategy based on scientific data. However, as the number of cougars increases, citing economic priority, hunting and ranching interests exert their political clout. They circumvent the desire of the general public and with pressure through an antiquated commission system are able to redirect the management efforts toward reducing cougars or preferably eliminating them again.

South Dakota appears to be immersed in this conflict between the desires of the public in general and special interest groups. Whether the people of the Black Hills and the rest of South Dakota resolve this conflict and make room for cougars depends on if they leave sufficient cougars alive to demonstrate that the only thing people have to fear from cougars is fear itself. If they do move beyond the current hysteria and adapt a sound ecologically based consensus on how to coexist with cougars, then there is hope for the rest of the Midwest. The Black Hills area is indeed a test case that is being watched by other midwestern areas. How it goes for cougars in the Black Hills will determine the fate of all cougars in the region. If it goes well, cougars can once again find an ecologically viable place in the great grassland ecosystem of the Midwest. If not, although biologically the Midwest could support viable populations of cougars, socially it will be a black hole, a population sink where cougars from the West come to eventually be hunted down and die in a blast of hysteria and gunfire. Only time will tell.

6 East beyond the Plains

ALTHOUGH THERE ARE SEVERAL BOOKS about cougars in the eastern part of the United States, a book on cougars moving into the Great Plains region has to include a discussion of these more eastern areas. Many of the cougars entering the plains regions are only passing through to the forested areas of the East. What awaits them there? Are they bypassing the open prairie regions for a better life in the East? Or will they survive their trek through the prairie gauntlet just to end up in a vast uninhabitable region where they will meet certain death? What, then, awaits them beyond the plains?

This chapter is not meant to be a detailed thesis of cougars in the East, but to highlight what may be in store for cougars successfully making the trip to eastern forests. Again we need to briefly look at the biological possibility of the East supporting cougars and then the sociological-political landscape that may await them. For the biological aspects, the question becomes: Is there enough habitat and are there enough deer to support viable populations of cougars?

THE HABITAT

Forested habitat is more conducive to the survival of cougars than are the open plains. The fact that cougars in the West and now those moving into the Great Plains seek out wooded areas attests to this affinity of animal and forest. Does the East offer enough forest habitat for cougars? The overall answer to that is a resounding yes. Although at one time most of the states east of the Mississippi River suffered extensive deforestation, much of the eastern landscape has reverted to forests. There still are extensive regions of open farmland in parts of the East, but less profitable lands, because of the success of the Great Plains, have been abandoned and in a process I refer to as defragmentation, much of the eastern forest is regrowing and reconnecting. I have seen this in all the states of the East I have traveled through, and one can easily see it from the satellite's-eye view of Google Earth. In response to this defragmentation, other wildlife species such as black bears and moose are also resurging in many parts of the East. It is expected that this process will also enable the return of cougars to this vast region.

Besides much of the eastern forest regrowing and reconnecting, there are large tracts of forested public land that can easily be homes to cougars. I have already mentioned the Ozarks in Missouri and Arkansas, but large national forests also await cougars—some think they have already arrived—in Minnesota, Wisconsin, and Michigan. Farther east there are the Appalachian Mountains with vast expanses of public and private forest extending up and down the eastern seaboard. Within that chain, areas such as the Adirondacks in upstate New York, which is bigger than Yellowstone National Park and a few others combined, provide more habitat than the Black Hills of South Dakota. If cougars can make it in the Black Hills, they can surely make it in these extensive areas in the East. So yes, without a doubt, there is ample habitat to support the return of cougars to the East. The next question is: Will they find enough food upon their arrival?

The Prey

There is no doubt that there are sufficient white-tailed deer to support cougars in the East. As in the prairie states, eastern states do not have a shortage of deer. This was not the case at one time; as with most wildlife resources, deer were overhunted to near extinction across most of the East. Due in large part to conservation-minded citizens, however, deer have rebounded. In fact, as expected, many feel there are too many deer, that their numbers and tameness are threatening the integrity of the recuperating forest.[1] Coyotes, who have filled the vacuum caused by the loss of wolves and cougars, do not pose a significant enough threat to keep deer in their ecological place. The return of the cougars to the East would have similar ecological cascading effects as upon their return to the midwestern areas. They will be a welcome sight to the battered and over-browsed plants and trees of the region.

Elk, too, are making a comeback. There are now several places in the East where elk, original inhabitants of eastern forests, have been reintroduced and are increasing in numbers. Whether these animals, like the deer, will rapidly reach population levels where they will bite the hand that feeds them remains to be seen. However, based on elk populations out West, the potential is surely there. Without the return of cougars and/or wolves to these areas, the reintroduction of elk may constitute further ecological suicide for eastern forests. Again, though not preyed on as much as deer, cougar predation on elk in the East can only help restore and maintain ecological balance with this large ungulate.

There appears to be plenty of prey waiting for cougars who arrive in the East, a virtual paradise waiting for the first feline Adam and Eve. And as predicted in the plains region, the arrival of the cougar in the East could have the benefit of mollifying the conflict between the health of the forest and the presence of large ungulates. Cougars, as in the plains, could become the shepherds of the eastern forest, keeping deer and elk in their places, and thus preserving the integrity and health of the eastern forest ecosystems. That is, if they can catch them. This leads us to the next consideration regarding cougars arriving in the East: There is plenty of cover and prey, but are the conditions right for cougars to catch sufficient prey to survive, and are there forest edges?

Hunting Habitat

It is one thing to have sufficient forest and prey for cougars; it is quite another to have the right conditions for cougars to catch their prey. As discussed in chapter 2, cougars are an edge species. They require edge or edgelike areas to see and then stealthily stalk their prey. Do the eastern forests provide this stalking habitat? In one of those many ecological paradoxes, the mature, closed precolonial eastern forest probably was not very good habitat for either deer or cougars. The survival of deer, and thus cougars, in those times probably depended on areas where indigenous people cultivated corn and other crops at the expense of forest cover. These areas, as they bordered the forest, were probably ideal for deer and their predators, including cougars. Modern farms in the East, where deer are most numerous and there is ample edge habitat, are the current equivalent to those precolonial croplands. Though indigenous people of the region apparently had no problem living among prey and predator, modern farmers would think differently about cougars. Unfortunately, then, the East, where these croplands are, is one of the least desirable places to allow cougars to exist.

Does this mean that the East, with all its abundant habitat, will not be able to support viable cougar populations? Fortunately, vast expenses of wilder areas in the East do not consist of forests with completely closed canopies. Natural events such as tornados, high winds, and even occasional forest fires set back the ecological clock to a younger, more open habitat in various parts of the eastern forest. There are also areas over much of the East that are being logged, again. These newly opened areas attract deer and will support cougars' efforts to catch a meal. The end result is that even a dense eastern forest consists of a crazy-quilt of closed and open areas. The makeup of this quilt shifts over time as new areas regrow and still newer ones open up. It is within this mobile collage that ancestral deer and cougars lived in the eastern forests and will do so again.

The fact that there still are cougars (Florida panthers) in the East attests to the ability of eastern forests to support the return of cougars to other regions. The East is truly a vacant forest waiting for its top predator to return. And upon its return, I predict it will be biologically and ecologically welcomed with open arms. How open the arms of the humans living in this vast region will be is another story.

The Human Factor

I predict that cougars in the East will face a similar sociopolitical environment as in the plains region. Easterners have not lived with cougars for several generations. There will also be the same unfounded concerns of safety and effects on wildlife (i.e., deer) populations. As in the Midwest, the occasional cougar sighting causes emotional, at times hasty, reactions by the public. For example, the sighting of a Black Hills cougar in western Connecticut sparked unfounded public concerns about being at risk of attack by this vicious wild beast. However, there are people, just as in the Midwest, who favor the cougar's return, and some of these groups (e.g., the Cougar Rewilding Foundation) are actively seeking the reintroduction of cougars into the East. Of the few attitude surveys that have been conducted, the most that can be said is that the general public is not adamantly *against* the return of the cougar to the East. In fact, many believe they are already/still there. Unfortunately, there is scant evidence that viable cougar populations in the East exist outside of southern Florida. How those attitudes may change when cougars start showing up consistently will depend a lot on what happens in the Midwest. It is hoped that by the time cougars return to the East, their record of behavior in the Midwest will mollify any concerns easterners may have regarding issues of safety.

The one factor working against the cougar's return will likely be the hunters. Long accustomed to a high hunting success rate, they will begin, as have their more western counterparts, to clamor for reductions or removals of cougars because of concerns they will threaten already too abundant deer herds. Similar arguments about who has the say about managing wildlife will arise as hunters try to assert their control over these public resources. Again, how these issues will be settled in the Midwest will set precedents for who eventually has the right to say how cougars in the East will be managed. And the resolution of this issue will determine the social and political landscape that will face relocated cougars. Fortunately, even if there is a high level of intolerance of cougars in the East, I predict that once cougars become established they will at least be able to live in the more remote areas in the East, for example northern Vermont, New Hampshire, and Maine. There, animals will survive with little influence by humans. I hope, for the cougar's sake, this prediction is accurate.

Epilogue

Throughout this book I have tried to present as factual of a data set on cougar behavior and ecology as existing science allows. I have presented it as a predator biologist and an ecologist. Though some may accuse me of not being objective in my assessment, I counter that my presentation is the only way to be objective in the case of the cougar. To present the cougar's role and place based on our historical and cultural misconceptions, or our fear and greed, would not be objective. To me, being objective is presenting what we know based upon logic and reason. Most of our commonly held "beliefs" regarding cougars are far from logic or reason, and I have come to understand and appreciate the role these unique animals have in ecosystems and even in the lives of their prey. Contrary to many who view predators as evil and savage, I see them as only making a living as they have evolutionarily learned how: they do this out of necessity to survive. In doing so, predators have earned themselves an essential place in the ecosystem, one we cannot, we should not, deny them. They may not regulate the numbers of their prey, but they do regulate their behavior, through fear. It is this behavioral regulation that maintains ecosystem integrity, ecosystem diversity, and the ordered flow of energy into and out of the community. Because of this, I objectively view predators such as cougars essential to ecosystem health—and all ecosystems need their predators. It is this basis that I have used in presenting my thoughts and ideas regarding cougars in the Midwest. I have looked at their role in the past prairie ecosystems in this light and I have used it as a guide in assessing where we should let them return to what remains of that ecosystem. I have investigated how we should react to their return, what adjustments we should make to provide them room. True, it is with the premise that cougars should return to the Midwest that I have based most of my discussions. I have done this because we as humans need predators, large and small. For the health sake of the ecosystems that support us, humans must make room for predators wherever we can. I hope that this book will help more people see that need, make that room. If this book alters the fate of at least one cougar's encounter with humans in its favor, I will have succeeded.

Notes

INTRODUCTION

1. Mader 1995.
2. Laundré et al. 2001; Laundré et al. 2010.
3. Laundré and Hernández 2003.
4. Hansen 1992; Danz 1999.
5. Logan and Sweaner 2001; Bolgiano and Roberts 2005.
6. Baron 2004.
7. Hornocker and Negri 2010.
8. Bolgiano and Roberts 2005.

CHAPTER 1. PRE-SETTLEMENT RECORDS AND THE DEMISE OF COUGARS ON THE GREAT PLAINS

1. Thwaites 1906, 22:346.
2. http://www.legendsofamerica.com/NA-Totems.html.
3. http://www.angelfire.com/ca/Indian/stories.html.
4. Morris 2002.
5. Gelb 1993.
6. Able 1970.
7. Thwaites 1906, vol. 13, vols. 22–24.
8. Thwaites 1904, vol. 27.
9. Thwaites 1904, vol. 17.
10. Thwaites 1904, vol. 13.
11. Thwaites 1904, vols. 22–24.
12. Clark et al. 2002.
13. Hoffmeister 1989.
14. Cory 1912.
15. Allen 1869, cited in Young and Goldman 1964.
16. Young and Goldman 1964.

17. Dinsmore 1994.
18. Ibid.
19. Ibid., 46.
20. Dinsmore 1994.
21. Young and Goldman 1946.
22. Gelb 1993.
23. Herrick 1802.
24. Watkins 1802, in Young and Goldman 1946.
25. Thwaites 1906, vol. 24.
26. Young and Goldman 1946.
27. Ibid.
28. Jones 1949.
29. Young and Goldman 1946.
30. North Dakota Game and Fish Department 2006.
31. Young and Goldman 1946.
32. Tyler and Anderson 1990.
33. Ibid.
34. Thwaites 1905, vol. 13.
35. Young and Goldman 1946.
36. Ibid.
37. Tyler and Anderson 1990.
38. Ibid.
39. Pike et al. 1997.
40. Young and Goldman 1964.
41. Ibid.
42. Jackson 1961.

Chapter 2. Ecology of Prairie Cougars

1. Hernández and Laundré 2005.
2. Laundré et al. 2001, 2010.
3. Laundré and Hernández 2003; Holmes and Laundré 2006.
4. Laundré et al. 2001; Hernández and Laundré 2005.
5. Laundré et al. 2001.
6. Ripple and Beschta 2003, 2004, 2007.
7. Ripple and Beschta 2006, 2007.
8. Ripple and Beschta 2007.
9. Ripple and Beschta 2006.
10. Logan and Sweanor 2000.
11. Laundré and Loxterman 2005.
12. Laundré 2005.
13. Altendorf et al. 2001; Hernández et al. 2005.
14. Laundré and Loxterman 2007.
15. Laundré et al. 2001.

16. Brown et al. 1999.
17. Laundré 2010.
18. Laundré and Loxterman 2007.
19. Laundré et al. 2007.
20. Ripple and Beschta 2007.

CHAPTER 3. THE FUTURE OF THE COUGAR IN THE MIDWEST

1. LaRue and Nielsen 2011.
2. Laundré 2005.
3. Thom and Wilson 1983.
4. Anderson 1970.
5. Nuzzo 1994.
6. Anderson 1970.
7. Curtis 1959; Finley 1976.
8. Baker and Whitman 1989.
9. Küchler 1994.
10. Küchler 1964.
11. Young and Goldman 1946.

CHAPTER 4. TO THE PRAIRIES AND BEYOND

1. LaRue and Nielsen 2008.
2. Ibid.
3. Ibid.

CHAPTER 5. CHALLENGES FACING THE NEW PIONEERS

1. Beier 1991.
2. National Highway Traffic Association.
3. Hubbard and Nielsen 2009.
4. Ibid.
5. Olliff and Caslick 2003.
6. http://www.cdc.gov.
7. Shaw 1977.
8. http://www.nass.usda.gov/.
9. http://www.dfg.ca.gov.
10. http://www.easterncougar.org.
11. Connecticut Department of Environmental Protection 2007.
12. Levy 2006.
13. Ballard et al. 2001.
14. Laundré et al. 2006.
15. Unsworth et al. 1999; Laundré et al. 2006.
16. Cougar Management Guidelines 2005.
17. South Dakota Mountain Lion Management Plan 2010–15.
18. Gigliotti 2002.

19. SD Mountain Lion Management Plan 2010–15, appendix table 13.
20. Ibid., appendix table 12.
21. Laundré et al. 2007.
22. http://www.blackhillslions.org.
23. Freyfogle and Goble 2009.

Chapter 6. East beyond the Plains

1. Rooney 2010; Levy 2006.

References

Abel, A. H. 1970. *Chardon's journal at Fort Clark, 1834–1839: Descriptive of life on the upper Missouri; Of a fur trader's experiences among the Mandans, Gros Ventres, and their neighbors.* Freeport, NY: Libraries Press.

Anderson, R. C. 1970. Prairies in the prairie state. *Transactions of the Illinois Academy of Science* 63:214–21.

Altendorf, K. B., J. W. Laundré, C. A. López González, and J. S. Brown. 2001. Assessing effects of predation risk on foraging behavior of mule deer. *Journal of Mammalogy* 82:430–39.

Ballard, W. B., D. Lutz, T. W. Keegan, L. H. Carpenter, and J. C. deVos Jr. 2001. Deer-predator relationships: A review of recent North American studies with emphasis on mule and black-tailed deer. *Wildlife Society Bulletin* 29:99–115.

Barker, W. T., and W. C. Whitman. 1989. *Vegetation of the Northern Great Plains.* North Dakota State University Agricultural Experiment Station Report 111. Fargo, ND: North Dakota State University.

Baron, D. 2004. *Beast in the Garden: A modern parable of man and nature.* New York: W.W. Norton.

Beier, P. 1991. Cougar attacks on humans in the United States and Canada. *Wildlife Society Bulletin* 19:403–12.

Bolgiano, C., and J. Roberts. 2005. *The eastern cougar historic accounts, scientific investigations, and new evidence.* Mechanicsburg, PA: Stackpole Books.

Brown, J. S., J. W. Laundré, and M. Gurung. 1999. The ecology of fear: Optimal foraging, game theory, and trophic interactions. *Journal of Mammalogy* 80:385–99.

Burt, W. H. 1943. Territoriality and home range concepts as applied to mammals. *Journal of Mammalogy* 24:346–52.

Clark, D. W., S. C. White, A. K. Bowers, L. D. Lucio, and G. A. Heidt. 2002. A survey of recent accounts of the mountain lion (*Puma concolor*) in Arkansas. *Southeastern Naturalist* 1:269–78.

Connecticut Department of Environmental Protection. 2007. *Managing urban deer in Connecticut: A guide for residents and communities.* 2nd ed. Hartford, CT: Bureau of Natural Resources. www.ct.gov/dep.

Connolly, G. E. 1978. Limiting factors and population regulation. Pages 245–85 in O. C. Wallmo, ed., *Mule and black-tailed deer of North America.* Lincoln: University of Nebraska Press.

Cory, C. B. 1912. *The mammals of Illinois and Wisconsin.* Publications of the Field Museum of Natural History 153; Zoological Series 11. Chicago: [Field Museum of Natural History].

Cougar Management Guidelines Working Group. 2005. *Cougar Management Guidelines.* Bainbridge Island, WA: WildFutures Press.

Curtis, J. T. 1959. *The vegetation of Wisconsin: An ordination of plant communities.* Madison: University of Wisconsin Press.

Danz, H. 1999. *Cougar!* Athens: Swallow Press/Ohio University Press.

Davenport, M. A., C. K. Nielsen, and J. C. Mangun. 2010. Attitudes toward mountain lion management in the Midwest: Implications for a potentially recolonizing large predator. *Human Dimensions of Wildlife* 15:373–88.

Dinsmore, J. J. 1994. *A country so full of game: The story of wildlife in Iowa.* Iowa City: University of Iowa Press.

Finley, R. W. 1976. *Original vegetation cover of Wisconsin.* Map compiled from U.S. General Land Office notes, University of Wisconsin Extension. St. Paul, MN: North Central Forest Experiment Station.

Freyfogle, E. T., and D. D. Goble. 2009. *Wildlife law: A primer.* Washington, DC: Island Press.

Gelb, N. 1993. *Jonathan Carver's travels through America, 1766–1768: An eighteenth-century explorer's account of uncharted America.* New York: Wiley and Sons.

Gigliotti, L. M. 2002. *Wildlife values and beliefs of South Dakota residents.* Report HD-10-02.AMS. Pierre: South Dakota Department of Game, Fish & Parks.

Hansen, K. 1992. *Cougar, the American lion.* Flagstaff: Northland Publishing Company.

Hernández, L., and J. W. Laundré. 2005. Foraging in the "landscape of fear" and its implications for habitat use and diet quality of elk *Cervus elaphus* and bison *Bison bison.* *Wildlife Biology* 11:215–20.

Hernández, L., J. W. Laundré, and M. Gurung. 2005. Use of camera traps to measure predation risk in a puma-mule deer system. *Wildlife Society Bulletin* 33:353–58.

Herrick, C. L. 1802. *Mammals of Minnesota.* Geology and Natural History Survey of Minnesota, Bulletin 7. Minneapolis: Harrison & Smith.

Hibbard, C. W. 1943. A check-list of Kansas mammals. *Transactions of the Kansas Academy of Science* 47:61–88.

Hoffmeister, D. F. 1989. *Mammals of Illinois.* Urbana: University of Illinois Press.

Holmes, B. R., and J. W. Laundré. 2006. Use of open, edge and forest areas by pumas *Puma concolor* in winter: Are pumas foraging optimally? *Wildlife Biology* 12:201–9.

Hornocker, M., and S. Negri, eds. 2010. *Cougar: Ecology and conservation.* Chicago: University of Chicago Press.

Hubbard, R. D., and C. K. Nielson. 2009. White-tailed deer attacking humans during the fawning season: A unique human-wildlife conflict on a university campus. *Human-Wildlife Conflicts* 3:129–35.

Jones, J. K., Jr. 1949. The occurrence of the mountain lion in Nebraska. *Journal of Mammalogy* 30:113.

Jackson, H. H. T. 1961. *Mammals of Wisconsin*. Madison: University of Wisconsin Press.

Johnsgard, P. A. 2003. *Lewis and Clark on the Great Plains: A natural history*. Lincoln: University of Nebraska Press.

Küchler, A. W. 1964. *Potential natural vegetation of the conterminous United States*. Special Publication no. 36. New York: American Geographical Society.

LaRue, M. A., and C. K. Nielsen. 2008. Modelling potential dispersal corridors for cougars in Midwestern North America using least-cost path methods. *Ecological Modelling* 212: 372–81.

———. 2011. Modelling potential habitat for cougars in Midwestern North America. *Ecological Modelling* 222:897–900.

Laundré, J. W. 2005. Puma energetic: A recalculation. *Journal of Wildlife Management* 9:723–32.

———. 2010. Behavioral response races, predator-prey shell games, ecology of fear, and patch use of a large predator and its ungulate prey. *Ecology* 91:2995–3007.

Laundré, J. W., and L. Hernández. 2003. Winter hunting habitat of pumas *Puma concolor* in northwestern Utah and southern Idaho, USA. *Wildlife Biology* 9:123–29.

Laundré, J. W., and J. Loxterman. 2005. Impact of edge habitat on summer home range size in female pumas. *American Midland Naturalist* 157:221–29.

Laundré, J. W., L. Hernández, and K. B. Altendorf. 2001. Wolves, elk, and bison: Reestablishing the "landscape of fear" in Yellowstone National Park, U.S.A. *Canadian Journal of Zoology* 79:1401–9.

Laundré, J. W., L. Hernández, and S. G. Clark. 2006. Impact of puma predation on the decline and recovery of a mule deer population in southeastern Idaho. *Canadian Journal of Zoology* 84:1555–65.

———. 2007. Numerical and demographic responses of pumas to changes in prey abundance: Testing current predictions. *Journal of Wildlife Management* 71:345–55.

Laundré, J. W., L. Hernández, and W. J. Ripple. 2010. The landscape of fear: Ecological implications of being afraid. *Open Ecology Journal* 3:1–7.

Levy, S. 2006. A plague of deer. *BioScience* 56:718–21.

Logan, K. A., and L. L. Sweanor. 2001. *Desert puma: Evolutionary ecology and conservation of an enduring carnivore*. Washington, DC: Island Press.

Mader, T. R. 1995. Mountain lion fact sheet. Unpublished report. http://www.aws.vcn.com/ mountain_lion_fact_sheet.html.

Morris, J. M. 2002. *Narrative of the Coronado expedition by Pedro de Castañeda de Nájera*. Chicago: R.R. Donnelley & Sons.

Muir, J. 1912. *The Yosemite*. New York: The Century Company.

Murphy, K., and T. K. Ruth. 2010. Diet and prey selection of a perfect predator. Pages 118–37 in M. Hornocker and S. Negri, eds., *Cougar: Ecology and Conservation*. Chicago: University of Chicago Press.

Nasatir, A. P. 1990. *Before Lewis and Clark: Documents illustrating the history of the Missouri, 1785–1804.* Vol. 1. Lincoln: University of Nebraska Press.

North Dakota Game & Fish Department. 2006. Status of mountain lions (*puma concolor*) in North Dakota. Unpublished report to the Legislative Council. North Dakota Game & Fish Department, Bismarck, North Dakota.

Nuzzo, V. A. 1985. Extent and status of Midwest oak savanna: Presettlement and 1985. *Natural Areas Journal* 6:6–36.

Olliff, T., and J. Caslick. 2003. Wildlife-human conflicts in Yellowstone: When animals and people get too close. *Yellowstone Science* 11, no. 1: 18–22.

Pike, J. R., J. H. Shaw, and D. M. Leslie Jr. 1997. The mountain lion in Oklahoma and surrounding states. *Proceedings of the Oklahoma Academy of Science* 77:39–42.

Ripple, W. J., and R. L. Beschta. 2003. Wolf reintroduction, predation risk, and cottonwood recovery in Yellowstone National Park. *Forest Ecology and Management* 184:299–313.

———. 2004. Wolves and the ecology of fear: Can predation risk structure ecosystems? *BioScience* 54:755–66.

———. 2006. Linking a cougar decline, trophic cascade, and catastrophic regime shift in Zion National Park. *Biological Conservation* 133:397–408.

———. 2007. Hardwood tree decline following large carnivore loss on the Great Plains, USA. *Frontiers in Ecology and the Environment* 5:241–46.

Rooney, T. P. 2010. What do we do with too many white-tailed deer? *ActionBioscience*, May. http://www.actionbioscience.org/biodiversity/rooney.html.

Samson, F. B., F. L. Knopf, and W. R. Ostlie. 2004. Great plains ecosystems: Past, present, and future. *Wildlife Society Bulletin* 32:6–15.

Shaw, H. G.1977. Impact of mountain lion on mule deer and cattle in northwestern Arizona. Pages 17–32 in R. L. Phillips and C. Jonkel, eds., *Proceedings of the 1975 Predator Symposium.* Missoula: Montana Forest and Conservation Experiment Station, University of Montana.

South Dakota Mountain Lion Management Plan. 2010–15. South Dakota Department of Game, Fish & Parks. Version 10-1, June 2010. http://mountainlion.org/States/South%20Dakota%20Draft%20Mountain%20Lion%20Plan%202010-2015.pdf.

Stolzenburg, W. 2008. *Where the wild things were: Life, death, and ecological wreckage in a land of vanishing predators.* New York: Bloomsbury.

Thom, R. H., and J. H. Wilson. 1983. The natural divisions of Missouri. *Natural Areas Journal* 3:44–51.

Thompson, D. J. 2009. Population demographics of cougars in the Black Hills: Survival, dispersal, morphometry, genetic structure, and associated interactions with density dependence. PhD diss., South Dakota State University.

Thwaites, R. G. 1904–6. *Early western travels, 1748–1846.* Vols. 13, 17, 22–24, 27. Cleveland, OH: Arthur H. Clark Company.

Tyler, J. D., and W. J. Anderson. 1990. Historical accounts of several large mammals in Oklahoma. *Proceedings of the Oklahoma Academy of Science* 70:51–55.

Unsworth, J. W., D. F. Pac, G. C. White, and R. M. Bartmann. 1999. Mule deer survival in Colorado, Idaho, and Montana. *Journal of Wildlife Management* 63:315–26.

Wasserberg, G., E. E. Osnas, R. E. Rolley, and M. D. Samuel. 2009. Host culling as an adaptive management tool for chronic wasting disease in white-tailed deer: A modelling study. *Journal of Applied Ecology* 46:457–66.

Young, S. P., and E. A. Goldman. 1964. *The puma, mysterious American cat*. New York: Dover Publications.

Index

Page numbers in bold indicate illustrations and captions.

prairies. *See* eastward movement across the prairies; Great Plains, generally

predators: attitudes toward, xi, 10; ecological role of, 9–10, 133–34, 177–78; food needs of, 32; landscape of opportunity for, 32–33, 35, 38–39, 47–49; lethality of, 32–34; predator–prey interactions/balance, 33–34, 155–56; weaknesses in hunting capability, 47

pre-settlement records and demise of cougars, 17–28; in Arkansas, 21; Chardon's account, 19; cougar's role in indigenous people's lives, 17; English accounts, 18; French accounts, 18; in Illinois, 21–22; in Iowa, 22; in Kansas, 22–23, 97; and the Lewis and Clark expedition, 19; Long's account, 20; in Minnesota, 23; in Missouri, 23; in Nebraska, 23–24, 90–91, 93; in North Dakota, 24, 82; Nuttall's account, 20, 24, 27; in Oklahoma, 24–25, 102; overview of, 13, 17–21; predators' elimination as a priority, 20–21; Prince Maximilian's account, 20, 23; river habitat of cougars, 27, 31; scarcity and secrecy of cougars suggested by, 26; in South Dakota, 25; Spanish accounts, 17–18; trappers'/settlers' accounts, 19–20; in Wisconsin, 25–26; vs. wolves and bison, 17–18, 26, 27, 31

Prince Maximilian of Wied, 20, 23

Pueblo (Colorado), 128–29

Pueblo Reservoir, 128

The Puma (Young and Goldman), 21–23, 25

Red Hills (Kansas), 100–101

Red River of the North, 76, 78–80, 113, 115

Red River of the South, 102, 127–28, 131–33

Redwood River, 77

reindeer, 150

Republican River, 91, 93, 124

riparian habitat: along Missouri River, 72, 79, 87; in Arkansas, 63, **64**, 65; damage to vegetation by grazers/browsers, 55–56; eastward movement via, 113, 115, 121, 123–24, 127, 129; ecological role of, 31–32, 35–39, 41, 48, 53, 56, 127; in Illinois, 68, **69**, 70; in Iowa, 72, **73**, 74; in Kansas, 97–98, **99**, 100–101; in Minnesota, **73**, 75–77, 75–78; in Missouri, **64**, 65–68; movement of cougars along, 54; in Nebraska, 90–91, **92**, 93–97; in North Dakota, 78–84, **81**, 82; in Oklahoma, **99**, 102–3; persistence of, 58, 60; river damming's effects on, 87; in South Dakota, **81**, 85–90, 86–89; in Wisconsin, 70–72

Ripple, Bill, 37, 55–56

risk assessment, 139

rivers: damming's effects on riparian habitat, 87; eastward movement via, 113, 115, 121–24, **125**, 127–28; ecological role of, 127; Midwest network of, 39, **40**, 41, 53–55, 58. *See also* riparian habitat; *and individual rivers*

Roberts, Thomas S., 23

Roosevelt, Theodore, 25

Rosebud tribal lands (South Dakota), 88

Ruth, Toni, 148

Salt River, 130

scientific method, 156–57

Scotts Bluff (Nebraska), 123

SDGFP (South Dakota Game, Fish & Parks) Department, 161–65, 168

sharks, 141

Shawnee National Forest (Illinois), 68, 70

sheep, 88–89, 147–49, 163

Sheyenne River, 79

shootings, deaths from, 140

shortgrass prairies, 11, 35, 90

Sioux City (Iowa), 122

Sitgreaves, Lorenzo, 25

smoking-related deaths, 140